OXFORD PAPERBACK REFERENCE

A GUIDE TO THE
Wild Orchids
of Great Britain
and Ireland

A GUIDE TO THE
Wild Orchids of Great Britain and Ireland

DAVID LANG

Second Edition

Oxford   New York
OXFORD UNIVERSITY PRESS
1989

Oxford University Press, Walton Street, Oxford OX2 6DP

Oxford New York Toronto
Delhi Bombay Calcutta Madras Karachi
Petaling Jaya Singapore Hong Kong Tokyo
Nairobi Dar es Salaam Cape Town
Melbourne Auckland

and associated companies in
Berlin Ibadan

Oxford is a trade mark of Oxford University Press

First published 1980 by Oxford University Press
Second Edition first published 1989 in the Oxford
Paperback Reference series

British Library Cataloguing in Publication Data

Lang, David, 1935–
A guide to the wild orchids of Great
Britain and Ireland.—2nd ed.
1. Great Britain. Orchids
I. Title
584'.15'0941
ISBN 0-19-282599-2

Library of Congress Cataloging in Publication Data

Lang, David, 1935–
A guide to the wild orchids of Great Britain
and Ireland / David Lang.—2nd ed.
p. cm.—(Oxford paperback reference)
Bibliography: p.
Rev. ed. of: Orchids of Britain. 1980.
Includes index.
1. Orchids—Great Britain—Identification. 2. Botany—Great Britain.
I. Lang, David, 1935– . Orchids of Britain. II. Title.
584'.150941—dc19 QK495.O64L28 1989 88-28990
ISBN 0-19-282599-2 (pbk.)

Typeset by CentraCet, Cambridge
Printed in Hong Kong

Contents

List of Species

The species number also refers to the colour plate (after page 118) and distribution map (after page 156).

1 Lady's-slipper   *Cypripedium calceolus*
2 White Helleborine   *Cephalanthera damasonium*
3 Narrow-leaved Helleborine   *Cephalanthera longifolia*
4 Red Helleborine   *Cephalanthera rubra*
5 Marsh Helleborine   *Epipactis palustris*
6 Broad-leaved Helleborine   *Epipactis helleborine*
7 Violet Helleborine   *Epipactis purpurata*
8 Slender-lipped Helleborine   *Epipactis leptochila*
9 Dune Helleborine   *Epipactis dunensis*
9a Young's Helleborine   *Epipactis youngiana*
10 Pendulous-flowered Helleborine   *Epipactis phyllanthes*
11 Dark-red Helleborine   *Epipactis atrorubens*
12 Ghost Orchid   *Epipogium aphyllum*
13 Autumn Lady's-tresses   *Spiranthes spiralis*
14 Summer Lady's-tresses   *Spiranthes aestivalis*
15 Irish Lady's-tresses   *Spiranthes romanzoffiana*
16 Common Twayblade   *Listera ovata*
17 Lesser Twayblade   *Listera cordata*
18 Bird's-nest Orchid   *Neottia nidus-avis*
19 Creeping Lady's-tresses   *Goodyera repens*
20 Bog Orchid   *Hammarbya paludosa*
21 Fen Orchid   *Liparis loeselii*
22 Coralroot Orchid   *Corallorhiza trifida*
23 Musk Orchid   *Herminium monorchis*
24 Frog Orchid   *Coeloglossum viride*
25 Fragrant Orchid   *Gymnadenia conopsea*
26 Small-white Orchid   *Pseudorchis albida*

Acknowledgements

Authors of experience are well aware of the lengthy period which must elapse between the conception of a book and the finished product. With this book the gestation period has been of truly elephantine proportions, covering no less than twenty-eight years.

Even now I can clearly remember finding an illustrated flora in the biology laboratory at Tonbridge School, being fascinated by the paintings of British orchids, and vowing one day to see them all.

My particular thanks are due to Francis Rose and the late Lynn Thomas who, in those early days and since, have guided my faltering botanical footsteps with never-failing courtesy and encouragement.

Over the years I have incurred a considerable debt of gratitude to botanical friends all over Great Britain and in Europe, many of whom have helped again in tracing notes and references needed in preparing the text. To all of them I offer my grateful thanks, especially to Peter Benoit, Mary Briggs, Michiel Ebbens, Lynne Farrell, the late Stuart Fawkes, Larch Garrad, Peter Lambley, John Lansley, John Lavendar, Jill Lucas, Erich Nelson, Kathleen Pickard-Smith, John Raven, David Streeter, and Hector Wilks.

During the early stages of preparing the text I received much help and encouragement from Joan Woods and from Nella Wilkinson who also prepared the entire typed draft.

I am particularly grateful to Peter Hunt and Franklyn Perring for checking the text, and for their advice and help. It is a pleasure for me, as an amateur within the discipline of botany, to record with thanks the generous help and advice I have always received from them and from their botanical colleagues.

To the Oxford University Press, and to Neil Curtis and Hilary Dickinson in particular, I offer my grateful thanks for their kindly guidance of a novice in the mystery of producing a book.

Finally, to my wife and children, who have, over many years in fair weather and in foul, climbed more hills and plodded through more bogs than most people have ever dreamed of, I can only say thank you, and dedicate this book to you and all who have suffered with a botanical father.

Lewes, Sussex   DAVID LANG
March 1979

Acknowledgements for the Second Edition

I am particularly grateful to M. R. Lowe for supplying one slide of *Dactylorhiza incarnata* ssp. *ochroleuca*, and the slides of *Dactylorhiza majalis* ssp. *cambrensis*, *D. majalis* ssp. *scotica*, and *Dactylorhiza lapponica*, and for help with redrafting the section on marsh-orchids.

My sincere thanks are also due to Dr F. Rose for considerable assistance in updating the distribution maps, and to Dr A. J. Richards for particular help with *Epipactis youngiana*.

Lewes, Sussex   DAVID LANG
June 1988

Introduction

The great attraction of wild orchids is their rarity and their beautiful and often strange appearance. Some species will disappear from a well-known site for years on end, only to reappear to the surprise and delight of the botanist searching more in faith than in hope.

The *Ophrys* species in particular are extraordinarily beautiful, combining rich colours with furry and velvety textures, the flower coming to resemble a bee, spider, or fly, with an uncanny degree of accuracy. Some species, such as the butterfly-orchids and the Fragrant Orchid, are richly scented, smelling of carnations, while the Lizard Orchid stinks of goats.

Because of the irregularity of their flowering appearance there is always the hope that one may come upon a field full of thousands of orchids.

Another delight in the study of orchids is the very complexity of their growth habits and lifecycles. So highly developed are some of their methods of multiplication that one wonders that they multiply at all. This very specialization carries with it an inherent risk of failure, so that many orchids which are on the extreme edge of their ecological range are very susceptible to climatic change or the lack of the proper insect vectors.

Since the first edition of this book in 1980 much has happened to alter the picture of orchids in the British Isles. Some species have been lost to vice-counties, while the range of other species has increased as they are found in new sites. Two very rare orchids, the Red Helleborine (*Cephalanthera rubra*) and the Ghost Orchid (*Epipogium aphyllum*), have been rediscovered in areas where they had not been seen for many years.

Two species new to the British Isles have been discovered—*Epipactis youngiana* in Northumberland and the Lapland Marsh-orchid (*Dactylorhiza lapponica*) in north-west Scotland (see pp. 140–3)—while recent research has put in doubt the nomenclature of several helleborines and marsh-orchids. The list of recorded hybrids has grown.

The form of Bee Orchid (*Ophrys apifera* ssp. *jurana*), also known as *O. apifera* var. *friburgensis* or var. *botteronii*, mentioned on p. 15, duly

I

turned up in 1984 in Wiltshire. *Ophrys bertolonii*, recorded in Dorset (see p. 152), proved to have been planted.

The following orchid species are all listed in the *Red Data Book*, and those marked with an asterisk enjoy special protection.

*Lady's-slipper	*Cypripedium calceolus*
*Red Helleborine	*Cephalanthera rubra*
Dune Helleborine	*Epipactis dunensis*
*Ghost Orchid	*Epipogium aphyllum*
Summer Lady's-tresses	*Spiranthes aestivalis*
Irish Lady's-tresses	*Spiranthes romanzoffiana*
Bog Orchid	*Hammarbya paludosa*
Fen Orchid	*Liparis loeselii*
Dense-flowered Orchid	*Neotinea maculata*
Late Spider-orchid	*Ophrys holoserica*
Early Spider-orchid	*Ophrys sphegodes*
Lizard Orchid	*Himantoglossum hircinum*
*Military Orchid	*Orchis militaris*
*Monkey Orchid	*Orchis simia*

Lewes, Sussex
June 1988

DAVID LANG

How to use the book

The first part of this book describes the structure and life history of the British Orchids, placing them in a general ecological context and explaining their classification. Each species is then described in detail illustrated by the relevant colour plates and, finally, there are the species distribution maps which indicate the areas of the British Isles where each grows.

The colour plates illustrate every orchid species which grows at the present time within the British Isles. The Summer Lady's-tresses is considered to be extinct in its classic sites in the New Forest. Additional colour plates show close-up details of each species, with any variants in form and colour. These photographs are the result of twenty-eight years of field study, and have all been taken in Great Britain.

No key to the identification of orchid species has been provided as it proved valueless to try to categorize all the characteristics of each species owing to their great variability.

The Structure of Orchid Flowers

At first sight the diversity of structure of the fifty species to be found in the British Isles is so great that no simple pattern is apparent. However, a closer look at the basic floral structure will show that they possess a great many features in common.

All the members of the family Orchidaceae belong to the class of Monocotyledons. This means that the young germinating plant has a single leaf, unlike say the familiar mustard of 'mustard and cress' which has a pair of first leaves.

They are all perennials with a fleshy root-stock or root tubers. The leaves are unstalked, undivided, and untoothed, often long and narrow, keeled, somewhat fleshy, and with parallel veins. In some of the saprophytic species the leaves are reduced to membranous sheaths or scales at the base of the flower stem.

The flowers are very variable, often showy in appearance, and borne in unbranched terminal spikes. The floral segments are not clearly divided into green sepals (calyx) and coloured petals (corolla), as is the case in so many flowering plants, but are often similar in structure. For this reason it is better to refer to them as perianth members, the six members being divided into two groups of three. The three outer perianth members form an approximate triangle, two lying laterally and one dorsally forming the upper part of the flower. The three inner perianth members consist of two lateral and one ventral, the ventral division being in most cases large and prominent, forming a structure called the labellum or lip.

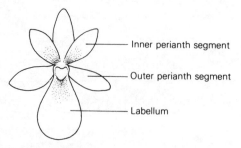

1 Diagram of a simple orchid flower

4

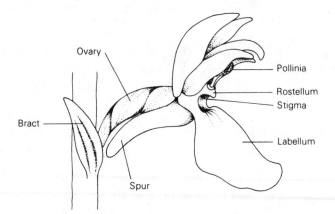

Ovary

Pollinia

Rostellum

Stigma

Bract

Labellum

Spur

2 Lateral view of a typical orchid flower

The labellum may be entire, lobed, or flat, the base being hollowed out to form a backwardly projecting spur which in some species contains nectar. In most species the labellum forms the lowest part of the flower, since in the developing bloom the ovary twists through 180°. However, in the Bog Orchid (*Hammarbya paludosa*) the labellum lies at the top of the flower, the whole having rotated through 360°, so that it comes to reoccupy its primitive position.

The male and female organs are borne on a common structure called the column. The stamens are reduced to one in all species, except the Lady's-slipper (*Cypripedium calceolus*) which has two, and each stamen is divided into two pollen-bearing masses called pollinia. The pollinia are formed of waxy, granular masses of pollen often borne on stalks called caudicles which are attached near the apex of the column. The stigmas number three in the *Cypripedium*. In other orchids there are one sterile and two fertile stigmas, the former reduced to a beak-like structure called the rostellum lying between the stamens and the fertile stigmas. In some species the rostellum bears one or two viscidia, sticky discs to which the pollinia are attached and which in turn attach the pollinia to visiting insects.

The ovary is either egg-shaped or cylindrical and lies between the base of the perianth members and the stem. Some species bear a small leaf-like bract at the point where the ovary joins the stem. The ovary is

5

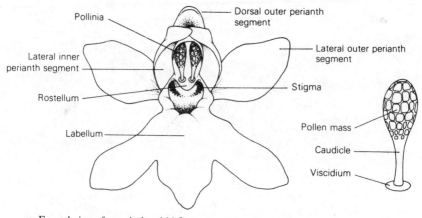

3 Frontal view of a typical orchid flower

composed of three united carpels and when ripe these open as three longitudinal valves, allowing the seeds to escape but leaving the three nerves of the carpels intact. The seeds are very numerous and dustlike, some 0.1–0.25 mm in length, rudimentary in structure, the embryo lying at the centre of an integumental veil with virtually no endosperm.

Germination

For the orchid seed to germinate the microclimate of the place where it lands must have the correct combination of moisture, oxygen, warmth, and light. However, it has been found that in many cases the seed must also become infected with a special fungus before germination can take place.

When the seed has germinated it forms a small peg-shaped structure called a protocorm. This has a bud at one end and proceeds to grow horizontally, producing rootlets at intervals. The next step is the production of tubers or roots and the growth upwards of a leaf-bearing stem. At this stage the protocorm withers, leaving the plant in its mature form.

Growth of the protocorm is very slow as the plant has no food reserves, and leaves may not be produced for several years. In the marsh-orchids, spotted-orchids, and the Green-winged Orchid the plant may reach maturity in four years while the same process in the Burnt Orchid (*Orchis ustulata*) may take fifteen years.

Mycorrhiza

The discovery of the role of mycorrhiza (root fungus) is credited to Noel Bernard (1874–1911). During experimental work with orchid seeds Bernard found that the tiny plants which had just germinated were infected with a fine fungal mycelium. No germination had occurred in those seeds placed in a sterile culture medium.

The mycorrhizal fungi, of which there are several species, belong mainly to the genus *Rhizoctonia*. They are not necessarily plant specific, and the same species of *Rhizoctonia* may be recovered from several orchid species. The fungus cannot be increased indefinitely in artificial culture, and in fact lessens its ability to cause germination of the orchid seed if grown separately for too long.

The fungus first attacks the orchid parasitically but is then checked and in its turn digested by the cells of the orchid. This alternating process continues all the time, although the dominance of each fluctuates, the orchid being dominant in the spring and summer, while the fungus is dominant in the autumn and winter when the plant's physiological processes are on a low level. The fungus converts substances like amidon generated by the breakdown of organic humus into soluble sugars which are more easily translocated in the young plant.

The dependence of the plant on its mycorrhizal fungus later in life varies with the orchid species involved. Those which grow large tubers tend to become independent and certainly very little fungal growth is detectable in these food-storage organs. Species such as the Red Helleborine (*Cephalanthera rubra*) and Creeping Lady's-tresses (*Goodyera repens*) with their extensive shallow rooting systems in surface humus, retain the mycorrhizal fungus throughout life.

Mycorrhiza exist in the roots of many forest trees such as beech, oak, birch, and pine, and in the roots of heathers. They can be roughly divided into those which live on the surface of the roots (ectotrophic) and those which live deep in the tissues of the roots (endotrophic).

7

This division is, however, artificial as all those which are surface growing show a degree of tissue penetration.

The Growth of Orchids

Orchids are perennials and have adapted to the inclement weather of winter by dying back each autumn to an underground tuber or fleshy root system, from which the fresh leaves emerge in the following spring, except in the Autumn Lady's-tresses (*Spiranthes spiralis*) and the Bee Orchid (*Ophrys apifera*), where the fresh leaves form in the autumn and persist throughout the winter. If you examine a flowering spike of the Autumn Lady's-tresses you will find a fresh crown of new leaves already formed beside it, from which next year's flowers will arise.

Many of the true orchids form egg-shaped or spherical tubers as food stores, while many of the helleborines store food in long fleshy roots which may be either horizontal or vertical. Each is beautifully adapted to the type of habitat in which the orchid grows. The round tuber of the *Orchis* species are well shaped to resist the desiccation which so often threatens the downland turf in which they grow. The shallow fleshy roots of the Irish Lady's-tresses (*Spiranthes romanzoffiana*) are well suited to the wet spongy terrain where it grows, while the long thin vertical roots of the Dark-red Helleborine (*Epipactis atrorubens*) can penetrate deep into the cooler moister crevices of the limestone rocks which it favours.

The Bog Orchid and Fen Orchid (*Liparis loeselii*) store their food in structures called pseudobulbs, which form at the base of the stem just above ground level.

Orchid plants may flower repeatedly—I have known one Frog Orchid plant which flowered for ten consecutive years. The Bee Orchid normally flowers but once, all its reserves being spent in one burst of flowering. Such a species is called monocarpic. All mature plants do not necessarily flower, this being especially true of the helleborines, and in large colonies of the Marsh Helleborine (*Epipactis palustris*) one can always find plenty of mature but non-flowering plants among those in flower.

Most orchids have green leaves, the cells of which contain chlorophyll, and they manufacture their food by normal photosynthesis, apart from that which they derive from their mycorrhiza. The Bird's-nest Orchid (*Neottia nidus-avis*), Coralroot Orchid (*Corallorhiza trifida*), and Ghost Orchid (*Epipogium aphyllum*) have reduced leaves and virtually no chlorophyll. All three have extensive underground systems which are heavily infected with both ecto- and endotrophic mycorrhizal fungi. In the Bird's-nest Orchid this consists of a dense mass of short fleshy roots like a badly built bird's nest. In the Ghost Orchid and the Coralroot Orchid there are no true orchid roots, but a greatly branched underground stem or rhizome. The Coralroot Orchid can grow in a more gritty soil containing less humus; but the Ghost Orchid can only grow in a mature beechwood where there is a considerable depth of rotted beech-leaf compost, through which the knobbly, coral-like rhizome can penetrate. The extent of this rhizome system is immense and I have seen rhizomes penetrating a rotted beech stump a good hundred metres from the nearest recorded flower spike.

Vegetative Multiplication

This process is to be found especially in woodland plants or those of high altitude where the lack of light or the low temperature does not favour the production of seeds.

In the *Orchis* and *Ophrys* species the plants may form two tubers in a really good year, these two tubers giving rise to two plants which in time become separate. This leads to the formation of clumps of plants and is frequently seen with marsh- and spotted-orchids. The plants of such a group will be identical in form, having derived from one parent, and are particularly conspicuous in populations which otherwise show considerable variation in the pattern of marks on the labellum.

In the Musk Orchid (*Herminium monorchis*) new tubers develop on the ends of stolons or runners, so that the new individuals which develop from them are well spaced out. In the Creeping Lady's-tresses and Ghost Orchid the extensive shallow rooting system gives rise to a series of individual flowering stems and there is very little seed production.

In the Common Twayblade (*Listera ovata*), Lesser Twayblade (*L. cordata*), and Red Helleborine the branching roots invariably develop buds from which new plants derive, so that the satellite plants often appear in lines stretching away from the parent plant. Such new plants may reach flowering maturity in three years or so, as opposed to the fifteen years normally taken from seed to flowering.

The Bog Orchid shows a very peculiar type of vegetative multiplication. Small fleshy knobs form on the edges of the leaves, detach, and grow separately. Mycorrhizal infection occurs at an early stage. The damp environment of the *Sphagnum* moss bogs where the plant grows favours this unusual type of propagation.

Sexual Reproduction

The various devices by which orchids are fertilized or ensure cross-pollination are among some of the most ingenious in the plant kingdom. Charles Darwin became fascinated by the manner in which the various orchids had adapted to the insect vectors which pollinated them, and in 1877 he published his famous paper *The Various Contrivances by which Orchids are Fertilised by Insects*. Thanks to the activities of the Kent Trust for Nature Conservation, Darwin's Bank, the downland area where many of his field observations were made, is now a nature reserve where the orchid species he recorded still flourish.

Although, as Darwin suggested, some orchids attract insects by a reward of nectar or sap, most do not, but operate an elaborate deceit mechanism to effect cross-pollination. Since the pollen is borne in sticky masses, it is never dispersed by wind, and self-fertilization is uncommon owing to the relative position of pollinia and stigma, separated in many cases by the rostellum.

If you examine the flower of the Early-purple Orchid (*Orchis mascula*) you will find that there are two pollinia, each with a caudicle and a separate viscidium. The viscidia are covered by a small flap called the bursicle, which prevents them from drying out prematurely. The flowers are visited by bees which feed on a fluid made within the fleshy wall of the spur.

As the bee's head pushes into the flower, it displaces the bursicle, and one or both of the viscidia stick on to its head or thorax. The sticky secretion at the base of the viscidium dries and hardens within

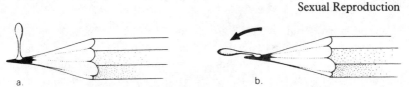

4 Movement of detached pollinia of the Early-purple Orchid, *Orchis mascula* (after Darwin)

thirty seconds so that it cannot easily be dislodged. Within the next minute a remarkable change takes place, the pollinia pivoting through 90° so that they now project forwards over the front of the bee's head. It is easy to demonstrate this experimentally by inserting a pencil into the orchid flower in such a manner that the pollinia are detached and stick to the point (Fig. 4). By this mechanism a time lapse is assured after the pollinia have been detached, during which it is likely that the feeding bee will have moved to another orchid plant. Only after the pollinia have pivoted forwards can they contact the stigma—before that they will only contact other pollinia, so no fertilization can occur.

A further refinement of this mechanism is seen in the long-spurred, nectar-bearing orchids such as the Pyramidal Orchid (*Anacamptis pyramidalis*) which are adapted to fertilization by moths or butterflies. The two pollinia arise from one viscidium, which is concave on its underside and whose edges roll in even further after detachment. This ensures a firm grip on the long, slender proboscis of the insect. The pollinia also swing forwards and outwards, thus ensuring contact with the stigmas which lie laterally.

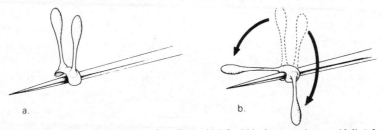

5 Movement of detached pollinia of the Pyramidal Orchid, *Anacamptis pyramidalis* (after Darwin)

A similar method is employed by the Greater Butterfly-orchid (*Platanthera chlorantha*) which is pollinated mainly by night-flying moths, attracted by the scent which is much stronger at night, the white flowers being easily detectable in the dim light.

In the Common Twayblade the column is short and the pollen very crumbly in texture. This tends to be deposited in a mass on the rostellum, which is gutter shaped. Nectar is secreted in a groove along the centre of the labellum. If the feeding insect, usually a small fly, touches the rostellum, a small drop of sticky fluid is explosively released, gluing the pollen mass to the insect's head and sending it in fright to another flower. Even if the fly visits other flowers on the same stem the pollen cannot be deposited on the stigma since, at that stage of development, the rostellum obstructs it. When the flowers are more mature the rostellum hinges up, exposing the stigma, which can then receive pollen from another plant, thus ensuring cross-pollination.

The Broad-leaved and Violet Helleborines (*Epipactis helleborine* and *E. purpurata*) also have very friable pollen masses which are easily broken by the wasps which visit them—the release is not, however, explosive.

Self-pollination occurs most frequently in those species where the rostellum is small. In the Slender-lipped Helleborine (*Epipactis leptochila*) the rostellum shrivels before the flowers are fully open, while in the Dune Helleborine (*E. dunensis*) and White Helleborine (*Cephalanthera damasonium*) there is virtually no rostellum, so that the pollen mass can fall of its own accord straight on to the stigma. The efficiency of such a mechanism is well demonstrated by the frequency of large fat seed capsules to be seen after a colony of White Helleborines has finished flowering.

Self-pollination also occurs with considerable frequency in the Bee Orchid, a flower highly adapted by mimicry for insect-pollination and endowed with a well-developed rostellum. The round yellow pollinia are borne on long caudicles which lie in grooves under the rostellum, which resembles a duck's head. The caudicles shrink as the flower matures, pulling the pollinia out of their protective pouches so that they swing down under their own weight straight on to the stigma. This can be seen very clearly in the close-up photographs of the Bee Orchid (Pl. 30).

The Lady's-slipper has a very different but no less ingenious method of ensuring fertilization. Insects are attracted to the nectar secreted inside the slipper-shaped lip, and land on the flat upper part of the column, called the staminode. They enter the slipper with ease but are prevented from climbing out by the curved and slippery inside wall. The only point of escape is near the base of the column where stiff hairs act as a climbing aid, but the insect has to squeeze through a relatively narrow exit, during which it rubs against the stamen collecting a load of pollen in the process. On visiting the next flower the pollen is brushed off against the stigma, which projects down into the slipper. This mechanism will only work for bees of the right size. Small insects will collect no pollen, while very large insects are incapable of escape from the slipper.

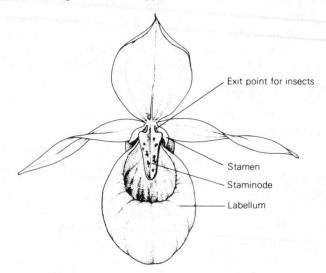

Exit point for insects

Stamen

Staminode

Labellum

6 Lady's-slipper, *Cypripedium calceolus* × 1½

One of the most bizarre methods of insect-pollination is seen among plants of the genus *Ophrys*, notably the Mirror Ophrys (*O. speculum*) of the eastern Mediterranean. The plant attracts the males of one species of insect only, probably in response to some secretion by the orchid, and from the insects' actions it is obvious that they are attempting to copulate with the flower. In so doing they remove pollinia on their abdomina and carry them to another flower. An interesting feature of this relationship is that the insects cease to visit the flowers when plenty of females of their own species are available.

Variation

Within most populations of plants of the same species individuals will be found which are not in every minute detail identical. Such subtle changes in colour or morphology represent the normal variation which occurs within a species.

In the Common Spotted-orchid (*Dactylorhiza fuchsii*), for example, there is considerable variation in the intensity and shape of the loops and dashes on the labellum. In the Green-winged Orchid (*Orchis morio*) a whole range of colours is recorded. While the majority of the flowers are a deep purple, others may be pale lilac, pink, white, or rarely a pale buff in colour. In certain species the degree of variation is so great, and occurs in whole populations, that it has been given taxonomic recognition, so that such plants have been granted varietal and even sub-specific rank.

The Early Marsh-orchid (*Dactylorhiza incarnata*) is, as its scientific name suggests, flesh pink in colour. Two other colour forms occur commonly and may even predominate in a geographical area. These colour forms are so distinctive that they have been given names—ssp. *pulchella* is mauve, while ssp. *coccinea* is a brilliant flaming red, especially when the flowers have just opened.

Both the Man Orchid (*Aceras anthropophorum*) and the Frog Orchid (*Coeloglossum viride*) combine green and reddish colours in their flowers. The proportions of these colours vary greatly, so that plants may have entirely greenish-yellow flowers or flowers of a dark red-brown colour. To some extent this depends on the amount of sunlight falling on the plant. Certainly Frog Orchids growing in shady places or in mountain areas tend to be greener, while one colony of Man Orchids I found growing in an area of exposed coastal shingle were a bright foxy brown.

Orchids with white flowers such as the butterfly-orchids show remarkably little variation in colour or shape.

Albinism occurs especially in those flowers whose basic colour is red, pink, or purple. The frequency of this varies greatly from species to species. White flowers are not uncommon in the Early-purple Orchid and the Fragrant Orchid (*Gymnadenia conopsea*), while I have only seen one albino Pyramidal Orchid in more than twenty years. Albinos also occur in the Common Spotted-orchid, but here one must sound a note of warning. A careful examination of many apparent albinos will reveal very faint marks on the labellum or sepals, while the pollinia are dark in colour. In the true albino the labellum is white, totally devoid of marks, and the pollinia are yellow.

In species such as the Bee Orchid where yellow pigments are present in addition to reds and purples, a condition of partial albinism occurs. For example, in *Ophrys apifera* var. *flavescens* (see Pl. 30) the outer perianth members are white, while the labellum is a curious sage green. A very similar colour variety occurs in the Marsh Helleborine (see Pl. 5).

In those species which bear spotted leaves that characteristic is often very variable, and within any group of plants individuals may have leaves unmarked, or, at the other extreme, heavily blotched. This is certainly true of both the spotted-orchids, the Northern Marsh-orchid (*Dactylorhiza purpurella*), and the Early-purple Orchid. An interesting form of leaf-spotting occurs in the Dense-flowered Orchid (*Neotinea maculata*). In some individuals lines of tiny reddish-purple spots are apparent on the leaves, such plants tending to bear flowers which are pink tinged rather than pure white.

Abnormal variation involves structural abnormality in some form, such variants on occasion being so weird and grotesque that the plant may be difficult to identify. Doubling of the floral parts may occur, so that the flower has a double column or a double lip. For example, a Fly Orchid (*Ophrys insectifera*) may be found where all the inner perianth members have become similar in structure, so that each 'fly' has another 'body' in the place of each 'antenna'.

I have seen Fragrant Orchids where all the perianth members have the structure of the outer perianth segments, and others where all the flowers were borne upside down. In one colony of Pyramidal Orchids many plants displayed entire unlobed lips. In the Bee Orchid and the Early Spider-orchid (*Ophrys sphegodes*) flowers occur where the various perianth segments are incompletely separated.

The Bee Orchid is renowned for the variants which occur. In var. *trollii* the labellum is long and pointed, barred with yellow and brown, while in the semi-peloric form the labellum resembles one of the outer perianth members. One curious form of the Bee Orchid, ssp. *jurana*, which is recorded in France and Holland, has the two upper inner perianth members similar in structure to the outer ones. It is a form which could easily be overlooked so that one should always check the structure of any Bee Orchid one finds. In 1986 this sub-species was found for the first time in Britain, in the south of England.

The flowering period of any species will show great variation depending on the part of the British Isles where it is recorded or the altitude of the site. However, in recent years a departure from normal flowering pattern has been recorded in the Burnt Orchid in East Sussex. In this area the Burnt Orchid normally flowers briefly in the last three weeks of May, setting seed and then withering with such rapidity that by mid-June it is an extremely difficult task to find any trace of a plant, even where sites have been tag-marked and the plant flowered in abundance.

In six sites plants have now been found flowering in the first two weeks of July, long after the earlier plants have vanished. Such flowers are morphologically indistinguishable from those flowering earlier, their slightly greater height probably being due to the relatively higher growth of the surrounding herbage at that time of the year. Some of the sites face north while others face south or south-west, so that incident light would not appear to play a part in the variation of the flowering period. In these sites no earlier-flowering plants have been recorded.

The Ecology of Orchids

Each species of orchid has its own special requirements which must be met for it to flourish, so that a study of the places where it is found will enable us to form an idea of the natural habitat of that species. Some are obviously woodland plants, others flourish in grassland of one sort or another, while others grow in specialized habitats such as marsh, bog, or fen.

Some plants are obviously more tolerant of variation in ecological conditions, so that they will be found in a number of different habitats, while others are exclusive and will only grow under very special conditions. Some habitats will not allow themselves to be neatly categorized but will have characteristics of more than one type, while orchids will sometimes be found where we least expect them to grow.

Mountain species will occur on downland, or woodland orchids in exposed sand-dunes.

However, with experience a certain pattern of occurrence does emerge, and the study of a site will enable one to predict with a fair degree of accuracy what orchid species will occur there. With these limitations in mind the following pages describe the types of habitat where orchids are found and the species occurring in those habitats.

Woodlands

BEECHWOOD

Beechwoods carry the richest orchid flora of all woodlands, which seems fitting for what are surely the most beautiful and majestic of British trees. They are mostly situated on chalk, often with a sheltered southern aspect, while the floor of the wood possesses a good depth of humus formed by the rotted beech leaves with only a small amount of competitive undergrowth.

The most characteristic orchid of the mature beechwood is the White Helleborine. It is tolerant of moderate shade, although it grows more vigorously on the edges of small clearings and will be found in thousands in early June in many beechwoods. Its rarer relative, the Narrow-leaved Helleborine (*Cephalanthera longifolia*), which comes into flower a week or so earlier, has a scattered distribution throughout the country and tends to flourish where the beech is not so mature, or the leaf canopy so thick. Rarely the two may be found growing together but it was only in 1973 that an apparent hybrid between the two was found in Hampshire.

Two orchids which are virtually restricted to mature beechwood are the Red Helleborine and the Ghost Orchid. The Red Helleborine will be found growing close up to the really large beech trees and obviously benefits from the lack of competitive undergrowth, but it is a shy flower in two of its sites and it would seem that a fair amount of light is necessary before it will bloom. Field-work is being undertaken at the present time to study the plant on the Continent, where it is a little less rare, so that we may determine the optimum conditions for it, thereby encouraging it to flower but not encouraging other plants.

Both the Bird's-nest Orchid and the Ghost Orchid grow in the depths of beechwoods where the light is often very dim. Since they have no leaves or chlorophyll-containing cells, but depend entirely on their mycorrhizal fungi, sunlight is of small importance. Also in the depths of the woods there will be the thickest layer of leaf compost so vital to their growth. Both have a basically pinkish-brown colour and are very hard to see against the background of dead beech leaves.

The Military Orchid (*Orchis militaris*) and the Lady Orchid (*Orchis purpurea*) also favour beechwoods, and the latter seems to have an affinity for yew trees. The classic sites for the Lady Orchid in Kent are mature beech 'hangers' on the slopes of the downs, with scattered, well-grown yews.

Along the edges of beechwoods, or in clearings where there is more light, there occurs a wealth of orchids. Broad-leaved Helleborine and Violet Helleborine are commonly to be found, especially on road verges where they may grow in showy clumps of as many as twenty spikes. Two rarer helleborines, the Slender-lipped and the Pendulous-flowered (*Epipactis phyllanthes*), also favour these conditions and in one colony of Pendulous-flowered Helleborines the plants grow up happily through a thick carpet of Ivy (*Hedera helix*) and Dog's Mercury (*Mercurialis perennis*). However, the light factor is relatively critical and in this same site the number of flowering spikes has decreased over the last twenty years from 121 in 1957 to 10 in 1975, as the trees have grown taller and thicker.

The Fly Orchid also tolerates the competition of Dog's Mercury very well, being found especially on the lower edges of beechwoods where less mature scrubby trees and bushes allow more light to fall.

In similar places both the Greater Butterfly-orchid and the Lesser Butterfly-orchid (*Platanthera bifolia*) are to be found, with plentiful Early-purple Orchid, Common Twayblade, and Common Spotted-orchid, the latter spreading out beyond the skirts of the wood into the surrounding grassland.

ASHWOOD

Ashwood on limestone is a woodland habitat found more especially in the north of the country, in Scotland and Yorkshire, but also in parts of Somerset. Although most of the trees are ash, there will usually be a fair proportion of hazel and oak and there are some fascinating woods of this type growing on limestone pavement, with areas of scree formed by the shattered limestone rock. The ground surface is surprisingly moist and has a well-developed covering of moss.

The most common orchids in these woods are the Early-purple Orchid, Common Spotted-orchid, and Common Twayblade. The Greater Butterfly-orchid and Fly Orchid are not uncommon, often appearing in large numbers when areas of the wood are felled, while the White and Broad-leaved Helleborines flourish where the trees are more mature and the ground cover less dense.

Woods of this type in the north of England are the habitat of the one known colony of the Lady's-slipper and also (i.e. it has many sites) of the Dark-red Helleborine. In these sites the wood is typically on a slope, with groups of trees separated by outcrops of limestone rock and grassy areas studded with Mountain Pansy (*Viola lutea*) and Bird's-eye Primrose (*Primula farinosa*).

OAKWOOD

Oakwood contains few orchid species and tends to be better after coppicing. Early-purple Orchids are relatively common, while less commonly Broad-leaved, Violet, and Pendulous-flowered Helleborines, Common Twayblade, Bird's-nest orchid, Greater Butterfly-orchid, and Common Spotted-orchid may all be found.

However, there are several very interesting oakwoods in north Wales and in Sutherland, where Narrow-leaved Helleborine occurs apparently quite naturally, the Sutherland locality consisting of oak and hazel growing by the sea.

PINEWOOD

New pine plantations are worthless places to search for orchids, but the old mature pine forests, especially of the Spey Valley in Scotland, have a fascinating flora which contains many orchids, one of which, the Creeping Lady's-tresses, is restricted to such habitat. There it flourishes in the moist, friable humus layer composed of moss and rotted pine needles, through which the fine root system spreads.

In clearings in the pinewoods the Lesser Twayblade will be found growing in the moss under the heather, while in the east of Scotland the Coralroot Orchid occurs locally in damp places in pinewoods with a fair amount of leaf mould or moss.

Along the edges of the rides and in more open spaces will be found the Early-purple Orchid, Heath Spotted-orchid (*Dactylorhiza maculata* ssp. *ericetorum*), Lesser Butterfly-orchid, and the Broad-leaved Helleborine. The Dune Helleborine grows in Lancashire and in Anglesey among sand-dunes on the edge of pine plantations where there is light scrub, and I have seen it growing several metres in among the closely ranked pine trees where there was still plenty of light.

BIRCHWOOD

The orchid flora of birchwoods is rather poor as so often they are situated on acid, leached soils. The commonest orchid is the Heath Spotted-orchid, while Common Twayblade, Broad-leaved Helleborine, and Common Spotted-orchid all occur, but rarely in large numbers.

ALDER CARR

These woods are formed by Common Alder (*Alnus glutinosa*) and various species of willow, growing in wet valley bottoms with poor drainage or slow-moving streams. They are also formed as fen areas dry out, when the Alders and willows become dominant.

Common Spotted-orchids and Common Twayblades are common, and in the south of England the Southern Marsh-orchid (*Dactylorhiza praetermissa*) occurs regularly, often growing to a considerable size. The large dense-flowering form of the Fragrant Orchid (*Gymnadenia conopsea* ssp. *densiflora*) flourishes in this habitat, flowering spikes often reaching a height of 1 metre, while in Scottish Alder carrs the Coralroot Orchid is sometimes to be found.

HAZEL COPPICE

The delicate tracery of the leaf cover of hazelwoods allows the sunlight to filter through, which leads to a flourishing population of orchids, flowers often appearing in considerable numbers following the cutting of the trees. Obviously the mature plants have lain dormant until the coppicing of the hazels allows more sunlight in, causing them to flower in the following spring and summer.

Early in the year, just as the Bluebells start to fade, the dark-purple spikes of the Early-purple Orchid appear. Common Spotted-orchids and Common Twayblades are plentiful, indeed in one Dorset wood the Common Twayblades flower so well that they comprise the dominant ground flora. Greater Butterfly-orchids and Fly Orchids are more frequently found here than in other woodlands, and in Kent one finds the Lady Orchid with huge shining leaves and superb flowering spikes of pink and brown flowers.

The Man Orchid may also be found in hazel coppice, although it is properly a grassland species and has usually spread into the wood from an adjacent field.

Grassland

DOWNLAND

Those of us who live in the south of England are fortunate to have the rolling grassland of the chalk downs to enjoy. Each area will have its own individual character, be it the North Downs of Kent and Surrey, the South Downs of Sussex, the Berkshire and Oxfordshire Downs, or the chalk hills of Wiltshire and Dorset. In addition to their intrinsic beauty, the downs are the home of many of our most attractive orchids, especially where the chalk turf is short and fine, the result of hundreds of years of grazing by sheep, undisturbed by the plough.

The fall in the sheep population dating from the First World War has resulted in the encroachment of downland by scrub and coarse tussock grass, while in the Second World War much ancient downland pasture was ploughed to grow desperately needed food. In recent years the advent of more efficient tractors and ploughs has enabled farmers to cultivate even further up the steep slopes of the downs, so that areas of ancient undisturbed downland turf are reduced to a fraction of their former acreage.

Very steep slopes have always defied cultivation, so that they, with the unploughed turf over prehistoric earthworks, barrows, and in long-abandoned chalk-pits are the most productive areas to search for orchids.

One of the first downland orchids to flower is the Early Spider-orchid on favoured downs in Kent, Sussex, and Dorset, often within sight of the sea. Soon afterwards the first spikes of the Early-purple Orchid appear, often in such masses that the slopes seem tinged with mauve. The Green-winged Orchid appears next, although it is getting much rarer nowadays, tending to favour the damper meadows near the foot of the downs where there is an admixture of clay in the soil. It also flourishes near the downs in churchyards and on garden lawns where it can grow undisturbed.

The Burnt Orchid flowers in late May at the same time that the Chalk Milkwort (*Polygala calcarea*) is at its peak, and from even a short distance it is very hard to differentiate between them. In the early part of June Common Twayblade, White Helleborine, and Greater Butterfly-orchid come into flower, especially where there is a little shelter from scrub as in the bottom of old chalk-pits, followed throughout June by the Common Spotted-orchid and the Fragrant Orchid. These last two species, together with the July-flowering Pyramidal Orchid, often flower in tens of thousands, carpeting the downs and scenting the air.

In high summer the Man Orchid, Monkey Orchid (*Orchis simia*), Lizard Orchid (*Himantoglossum hircinum*), and Bee Orchid are to be found flowering from mid-June well into July, when the Musk Orchid, Frog Orchid, and Late Spider-orchid (*Ophrys holoserica*) start to come into bloom. None of these can really be rated as of common occurrence, the Monkey Orchid and Late Spider-orchid being very rare and restricted to a handful of sites.

Latest of all the downland orchids is the Autumn Lady's-tresses, flowering in August and September. In some years it appears in thousands where the grass is really short. Like the Green-winged Orchid it flourishes on lawns, and I have also seen it in profusion on the tops of sea cliffs in Cornwall, growing in the short turf, often with Autumn Squill (*Scilla autumnalis*).

LIMESTONE GRASSLAND

The distribution of limestone grasslands is northerly, in the Pennines of Yorkshire, Lancashire, Derbyshire, and Durham, and in the Lake District. Many of the downland species are represented here, Early-purple Orchid, Burnt Orchid, and Fragrant Orchid, Lesser Butterfly-orchid (here replacing the Greater Butterfly), Fly Orchid, Frog Orchid, and the delightful Small-white Orchid (*Pseudorchis albida*) whose distribution is distinctly northerly, apart from limited areas of limestone grassland in Wales.

Another specialized northerly habitat is the machair, a type of calcareous grassland formed on stabilized shell sand. The finest stretches of machair are to be found in the Hebrides on the Isle of

Lewis and on Tiree, although less extensive areas exist on many other Hebridian islands and in limited coastal stretches of the north-west mainland of Scotland. The fine turf contains many of the downland flowers familiar to the southern botanist—Thyme (*Thymus* ssp.), Eyebright (*Euphrasia* spp.), White Clover (*Trifolium repens*)—the whole sward being closely grazed by sheep and cattle under carefully controlled conditions. Excessive grazing is quickly followed by erosion of the shallow turf and blow-out of the underlying sand in the high winds which are such a feature of the area.

The gem of the machair is certainly the Hebridean sub-species of the Common Spotted-orchid, *Dactylorhiza fuchsii* ssp. *hebridensis*, which is a dwarfed form with a strongly marked, three-lobed labellum and compact, rounded inflorescence. In June it covers the machair in countless numbers, with here and there tiny plants of the Frog Orchid and in the more sheltered dune areas Lesser Butterfly-orchids and Pyramidal Orchids.

LIMESTONE PAVEMENT AND CLIFFS

Limestone pavement is one of the most fascinating habitats, both botanically and scenically. Here the strata of limestone rock occur in flat terraces which are intersected with deep fissures formed by the erosion of softer material. The scene resembles a vast crazypaving, the fissures sometimes reaching a depth of several metres, posing a threat of fractured bones to the unwary. In these clefts one finds lime-loving plants, which flourish in the sheltered and surprisingly moist humus which has slowly accumulated.

Areas of limestone pavement or limestone cliffs are found in the North Yorkshire Dales, Cumberland and Westmorland, Teesdale in the Pennines, Great Orme's Head in North Wales, Kishorn, Durness, and Inchnadamph in Scotland, and the Burren in western Ireland. Early-purple Orchids are common in early spring, followed by the Common Spotted-orchid and Bee Orchid (except in Scotland). One of the specialities of the Burren is the Dense-flowered Orchid, a plant of principally Mediterranean distribution which flourishes along with the Spring Gentian (*Gentiana verna*), Shrubby Cinquefoil (*Potentilla fructicosa*), and many other delights in that strange lunar landscape.

Broad-leaved Helleborine and later the Autumn Lady's-tresses are frequently found on limestone pavement, but without a doubt one of the most attractive orchids to be found is the Dark-red Helleborine. The deep wine-red spikes of flowers are most impressive contrasting with the bleached white rock, often growing in inaccessible fissures in the cliff face, as on the Great Orme.

HILL MEADOWS

The small hayfields of the uplands of the north of England, Wales, and Scotland contain a wealth of orchids. These fields only yield a single cut of hay, usually in July, their small size and frequent dampness precluding the use of heavy farm machinery. They are mostly permanent pastures, since it is difficult to establish a worthwhile short-term ley where the soil is shallow and subject to severe weathering. Fortunately for the orchids the cost of artificial nitrogenous fertilizers makes their use rather uneconomic, so that the plants are able to flower relatively undisturbed, setting seed before the hay is cut.

The variety of orchid species to be found is truly astonishing. One such hill meadow near Dolgellau in Gwynedd contained Early-purple Orchids, Common and Heath Spotted-orchids, Common Twayblades, Early Marsh-orchids (*Dactylorhiza incarnata* ssp. *pulchella*), Northern Marsh-orchids, Fragrant Orchids, and both Greater and Lesser Butterfly-orchids. All were growing in profusion. Similar hill meadows in Scotland contain much the same species, with in addition the Small-white Orchid.

HEATH AND MOORLAND

Much heath and moorland is botanically rather monotonous and uninteresting, except where neutral or alkaline flushes occur. There one finds Early Marsh-orchids, Fragrant Orchids, and Lesser Butterfly-orchids. Over the remaining area the only orchid found with any frequency is the Heath Spotted-orchid.

In sheltered rocky areas where there is a good layer of *Sphagnum* moss underlying the Ling (*Calluna vulgaris*) and Bell Heather (*Erica cinerea*) the Lesser Twayblade may be found. It is a tiny plant, less than 5 centimetres in height and easily overlooked. The task of finding it is complicated by the resemblance of the non-flowering plant to a pair of Bilberry leaves (*Vaccinium myrtillus*). I have also found the Lesser Twayblade growing in the moss *Polytrichum* under scrubby Juniper bushes high up on the Yorkshire fells.

FEN

Fen is formed where a layer of peat is covered by neutral or alkaline water, the latter deriving from soluble chalky rocks. The most extensive fenlands are to be found in East Anglia. Each year drainage works and the slow tilting of the land mass are lowering the water table of the area, so that the fens are drying out.

To some people fens may appear lonely and desolate, but they have a great attraction for the naturalist on account of the rare plants and insects to be found there, indeed for some of these fen is their only habitat.

The Fen Orchid in its normal form is restricted to fen sites, although there is a form, *Liparis loeselii* var. *ovata*, which occurs in dune slacks in South Wales. It favours the areas where the reed *Juncus subnodulosus* has recently been cut back, so that there is only a moderate regrowth and a plentiful ground cover of moss. As the reeds grow up and thicken, so the number of orchid spikes is reduced, so that the traditional cropping of reeds is an integral part of the ecological pattern essential for the welfare of the orchids.

The Early Marsh-orchid flowers here in its normal colour and also in an unusual pale-yellow form *Dactylorhiza incarnata* ssp. *ochroleuca*, which is peculiar to the fens. The Southern Marsh-orchid is common and hybridizes freely with the rarer Irish Marsh-orchid (*Dactylorhiza traunsteineri*), the latter being restricted to fen sites. The hybrid seems to be stronger-growing than *D. traunsteineri*, flowering to the detriment of the latter as the fenland dries out. At one site in Suffolk over the last fifteen years the hybrid has entirely replaced *D. traunsteineri*. The Marsh Helleborine is a relatively common plant in the fens, flowering in late summer.

BOG

Bogs form where acid water lies over a peat base. Such habitat typically has a good covering of *Sphagnum* moss and Bog Asphodel (*Narthecium ossifragum*). The orchid species to be found here are few in number, but none the less very interesting. Along the drier margins and on raised humps in the bogs you will find the Heath Spotted-orchid and the Lesser Butterfly-orchid. In bogs in the north and west the Lesser Twayblade occurs where these humps are a little drier still, so that Bell Heather and Ling grow with the *Sphagnum* under them.

The Bog Orchid is without doubt the star attraction of the bogland orchids. It grows in well-developed sheets of *Sphagnum*, often along the edges of the main stands of *Phragmites* reeds, and it is not only rare but tiny, elusive, and remarkably erratic in its numbers from year to year. Less rare in such situations are the horseflies and I have a vivid memory of being driven from a superb colony of Bog Orchids in the New Forest by a bloodthirsty swarm of clegs.

MARSH

Here the soil contains only a little peat, quite a proportion of rotting vegetation, and a slow-moving or static sheet of water of fairly neutral *p*H. There is usually a profuse growth of reeds and rushes, with moss growing in places between them. This is truly the domain of the marsh-orchids, the species found varying according to the part of the country described. The Early Marsh-orchid is ubiquitous, the Southern Marsh-orchid predominates in the south and east, while the Northern Marsh-orchid is mostly to be seen in the marshes of Wales, Scotland, and the north of England. Recent field work has led to the discovery of a new marsh-orchid in north-west Scotland, and the necessary reclassification of the marsh-orchids there. The Broad-leaved Marsh-orchid *Dactylorhiza majalis* ssp. *occidentalis* is restricted to western Ireland, and the sub-species *cambrensis* to Wales. The whole situation is further complicated by the ease with which the marsh- and spotted-orchids hybridize with one another.

Both the spotted-orchids are common in marshland. The Irish Lady's-tresses in its classic sites in Ireland, such as the shores of Lough Neagh, grows in marshy areas where streams run into the lake. The Marsh Helleborine is to be seen at its best in marshes, sometimes flowering in such enormous numbers that one can scarcely walk without treading on the plants.

DUNES

Botanizing in established sand-dunes is a delightful pastime. There is no mud, it is within sight and sound of the sea, and the flora is always fascinating, combining seaside plants with those one associates more commonly with chalk pastures. The calcareous nature of many sand-dunes derives from the finely divided shell which forms a proportion of the sand itself.

Orchids tend to flourish in those areas of the dunes where the sand has been well stabilized and a fine covering of turf has developed, such areas in Scotland often being extensions of the machair already described. This is one reason why seaside golf-courses are so interesting botanically, although it is as well to botanize with discretion and a wary eye for a hard-driven ball.

Dune slacks, the moist low-lying areas between the hummocks of the stabilized dunes, are particularly worthy of investigation, the places being well marked in many cases by the extensive growth of Creeping Willow (*Salix repens*).

The Green-winged Orchid and the Early-purple Orchid are the first orchids to flower in the dunes, followed in early June by the Early Marsh-orchid. The dune form of this, *Dactylorhiza incarnata* ssp. *coccinea*, is of a brilliant red colour, most marked when it has just come into flower, and the sight of thousands of bright-red spikes dotting the turf of the dunes is breath-taking.

Throughout June the Common Spotted-orchid is in flower, replaced in the Hebrides by the short-stemmed, round-headed form *Dactylorhiza fuchsii* ssp. *hebridensis*. The Common Twayblade, the Southern Marsh-orchid, and the Northern Marsh-orchid are all abundant in favoured dunes, the two marsh-orchid species occurring together in several sites in north-west Wales. The Lesser Butterfly-orchid also flowers in June, followed by the Bee Orchid and the Pyramidal Orchid in more southerly localities. I have a vivid memory of a Cornish golf-links smothered in Pyramidal Orchids, the pink colour being particularly rich and dark near the sea.

Many rare species have duneland sites. The Lizard Orchid occurs regularly in two duneland areas of south-east England, while in South Wales the broad-leaved form of the Fen Orchid (*Liparis loeslii* var. *ovata*) flowers in the moister dune slacks among the Creeping Willow.

The Marsh Helleborine grows in abundance in wet dune slacks especially in the north and west of the country, while the Pendulous-flowered Helleborine has an outpost in North Wales, and the Dune Helleborine has its classic dune sites in Anglesey and Lancashire.

On the east coast of Scotland the Coralroot Orchid has its main distribution centre in sand-dunes, accompanied in areas where there are pines and heather by the Creeping Lady's-tresses and the Lesser Twayblade.

In recent years the Dense-flowered Orchid has been found in dunes in the Isle of Man.

Duneland is truly a fascinating habitat, but one very susceptible to erosion by weather and the pressure of human activity, something we too easily forget.

Classification

The classification and identification of orchids found in this country has often proved difficult for the amateur botanist, partly because the floral parts of the orchids themselves are so peculiar and complex that a detailed description in words alone does not create a clear enough picture for the enquirer. This situation is further complicated by the many changes of nomenclature which have been made over the years, as our knowledge of the plants has increased. It is indeed puzzling to find a plant moving from genus to genus, trailing in its wake a string of different scientific names.

The chart on pages 30–1 sets out the orchids occurring in Great Britain in their families, tribes, and genera. A brief description of the characteristics of each genus follows, so that an unknown plant can be tracked to its correct genus.

The detailed description of each species is included in the text for that species, the plants being described in the same order as laid out in the chart. This order of genera and species is based on the most widely accepted classification of the family, and is that followed by Clapham, Tutin, and Warburg, and by Perring and Walters in the *Atlas of the British Flora*.

The scientific synonyms for each species are printed following the presently accepted nomenclature, so that there should be no confusion regarding plants previously well known by other names. Indeed the curious situation arises when talking of orchids that there is often a single well-known English name, but a plethora of scientific relations.

All the flowers classified under the family Orchidaceae are perennial with rhizomes, vertical roots, or tubers. All are dependent at one stage in their life on a mycorrhizal fungus, and some are wholly or partially saprophytic.

The leaves are entire, often sheathing the stem, and may bear dark spots or blotches. In the saprophytic species the leaves may be reduced to membranous sheaths or scales.

The inflorescence is either a spike (a conical flowering head with sessile flowers) or a raceme (a conical flowering head with flowers borne on short stalks called pedicels).

The flowers have one plane of symmetry, an inferior ovary, and are usually bisexual. The six floral segments are borne in two whorls of three, the posterior segment of the inner whorl—the labellum—being in most cases large, complex, and highly coloured. In the majority of species the labellum points downwards owing to the twisting through 180° of the ovary or its pedicel. The base of the labellum commonly bears a backwardly projecting hollow spur. The anthers and stigma are borne on a special structure called the column.

The family Orchidaceae is divided into two sub-families:

Sub-family A Diandrae

The stamens are two in number, carried one on either side of the column, the third primeval stamen being changed into a petalloid structure—the staminode. Only one genus is represented in Great Britain.

Genus *Cypripedium*
Flowers very large, usually solitary. The dorsal sepal, conjoined ventral sepals, and two lateral petals, all purplish brown in colour, are set in the form of a cross. The labellum is very large and inflated, in the shape of a yellow slipper.

Sub-family B Monandrae

These possess one stamen occupying a more or less central position. This sub-family is further divided into the division Acrotonae where the stamen is attached by its tip to the column, and the division Basitonae where it is attached by its base.

Division A Acrotonae

In the tribe Neottieae the stamen is found at the back of the column—this includes all the helleborines, lady's-tresses, and twayblades. In the tribe Epidendreae—Fen and Bog Orchids—the stamen is borne near the tip of the column.

Family — Orchidaceae

SUB-FAMILY A Diandrae

TRIBE
Cypripedieae

1 *Cypripedium calceolus* L. — Lady's-slipper

SUB-FAMILY B Monandrae

Division A Acrotonae

TRIBE   **SUB-TRIBE**
Neottieae   i Cephalantherinae

2 *Cephalanthera damasonium* (Miller) Druce — White Helleborine
3 *C. longifolia* (L.) Fritsch — Narrow-leaved Helleborine
4 *C. rubra* (L.) Richard — Red Helleborine

ii Epipactinae

5 *Epipactis palustris* (L.) Crantz — Marsh Helleborine
6 *E. helleborine* (L.) Crantz — Broad-leaved Helleborine
7 *E. purpurata* Smith — Violet Helleborine
8 *E. leptochila* (Godfery) Godfery — Slender-lipped Helleborine
9 *E. dunensis* (T. & T. A. Stephenson) Godfery — Dune Helleborine
9a *Epipactis youngiana* A. J. Richards & A. F. Porter — Young's Helleborine
10 *E. phyllanthes* G. E. Smith — Pendulous-flowered Helleborine
11 *E. atrorubens* (Hoffman) Schultes — Dark-red Helleborine

iii Epipogiinae

12 *Epipogium aphyllum* Swartz — Ghost Orchid

iv Spiranthinae

13 *Spiranthes spiralis* (L.) Chevallier — Autumn Lady's-tresses
14 *S. aestivalis* (Poiret) Richard — Summer Lady's-tresses
15 *S. romanzoffiana* Chamisso — Irish Lady's-tresses

v Listerinae

16 *Listera ovata* (L.) R. Brown — Common Twayblade
17 *L. cordata* (L.) R. Brown — Lesser Twayblade
18 *Neottia nidus-avis* (L.) Richard — Bird's-nest Orchid

vi Physurinae

19 *Goodyera repens* (L.) R. Brown — Creeping Lady's-tresses

Epidendreae   Liparidinae

20	*Hammarbya paludosa* (L.) Kuntze	Bog Orchid
21	*Liparis loeselii* (L.) Richard	Fen Orchid

Vandeae   Corallorhizinae

22	*Corallorhiza trifida* Chatelin	Coralroot Orchid

Division B Basitonae

TRIBE   SUB-TRIBE
Ophrydeae   i Gymnadeniinae

23	*Herminium monorchis* (L.) R. Brown	Musk Orchid
24	*Coeloglossum viride* (L.) Hartman	Frog Orchid
25	*Gymnadenia conopsea* (L.) R. Brown	Fragrant Orchid
26	*Pseudorchis albida* (L.) A. & D. Löve	Small-white Orchid
27	*Platanthera chlorantha* (Custer) Reichenbach	Greater Butterfly-orchid
28	*P. bifolia* (L.) Richard	Lesser Butterfly-orchid

ii Serapiadinae

29	*Neotinea maculata* (Desf.) Stearn	Dense-flowered Orchid
30	*Ophrys apifera* Hudson	Bee Orchid
31	*O. holoserica* (Burm. fil.) W. Greuter	Late Spider-orchid
32	*O. sphegodes* Miller	Early Spider-orchid
33	*O. insectifera* L.	Fly Orchid
34	*Himantoglossum hircinum* (L.) Sprengel	Lizard Orchid
35	*Orchis purpurea* Hudson	Lady Orchid
36	*O. militaris* L.	Military Orchid
37	*O. simia* Lamarck	Monkey Orchid
38	*O. ustulata* L.	Burnt Orchid
39	*O. morio* L.	Green-winged Orchid
40	*O. mascula* L.	Early-purple Orchid
41	*Dactylorhiza fuchsii* (Druce) Soó	Common Spotted-orchid
42	*D. maculata* (L.) Soó ssp. *ericetorum* (E. F. Linton) Hunt & Summerhayes	Heath Spotted-orchid
43	*D. incarnata* (L.) Soó	Early Marsh-orchid
44	*D. praetermissa* (Druce) Soó	Southern Marsh-orchid
45	*D. purpurella* (T. & T. A. Stephenson) Soó	Northern Marsh-orchid
46	*D. majalis* (Reichenbach) Hunt & Summerhayes	Broad-leaved Marsh-orchid
47	*D. traunsteineri* (Sauter) Soó	Irish Marsh-orchid
47a	*Dactylorhiza lapponica* (Laest. ex Hartman) Soó	Lapland Marsh-orchid
48	*Aceras anthropophorum* (L.) Aitken. f.	Man Orchid
49	*Anacamptis pyramidalis* (L.) Richard	Pyramidal Orchid

Genus *Cephalanthera*

Shortly rhizomatous plants with leafy stems and lax spikes, with a few large sub-erect white or pink sessile flowers which never open widely. The labellum has a cup-shaped hypochile at its base which clasps the column. There is a constriction at the middle of the labellum, and the distal portion, the epichile, is forwardly directed, ridged on its upper surface, and has a recurved tip. There is no spur.

Genus *Epipactis*

Rhizomatous plants with leafy stems which bear sheathing scales at the base. The largest leaves are near the middle of the stem, a few shorter broader leaves below, the narrower leaves above grading into the bracts. The flowers are rather inconspicuous in a one-sided raceme. The labellum is again divided into hypochile, which is non-clasping, and epichile, which is triangular and downward pointing. It bears no ridges but has two bosses at its base. There is no spur. The column is short and the ovary straight.

Genus *Epipogium*

There is only one representative of this genus, *Epipogium aphyllum*. The plant arises from a complex root system resembling white coral. It is saprophytic, bears no leaves, and the stem is swollen just above the base. The labellum which is large, triangular, and pink is directed upwards, as is the fairly long spur.

Genus *Spiranthes*

Small plants with two to six rather tuberous roots, leafy stems, and flowers in a spirally twisted spike. The flower is in the form of a two-lipped tube around the column, the upper lip formed by the adherent perianth segments and the lower by the labellum. This is frilled, furrowed below, and slightly recurved at the tip. There is no spur.

Genus *Listera*

Plants with short rhizomes, erect stems, and a pair of opposite leaves borne some way up the stem. Flowers short stalked in a rather lax raceme. Apart from the labellum, the perianth segments are of similar length. The labellum bears two tiny lateral lobes, the main lobe being boldly forked almost to half-way. There is no spur. The capsule is globular.

Genus *Neottia*

Saprophytic plants containing no chlorophyll, with a dense mass of fleshy roots like a tangled bird's nest. The stem is leafless and covered with brownish scales. Brownish flowers in a moderately dense spike, the perianth segments curved to form a hood above the long broad labellum, which is divided at its tip into two diverging lobes. There is no spur.

Genus *Goodyera*

There is no tuber but a fleshy stoloniferous rhizome. The leaves are ovate, stalked, and bear a network of prominent veins. The flowers are small and white, carried in a spiral row, the outer perianth segments somewhat spreading, the labellum with a narrow spout-like distal portion. All floral parts are covered with glandular hairs. There is no spur and the bracts are longer than the ovary.

Genus *Hammarbya*

Small green plants always growing in *Sphagnum* moss. Stem with a basal pseudobulb, oval concave leaves, with tiny green bulbils on the margins, borne on the middle of the stem, below which are one or two leaves reduced to sheaths. The flowers are small and green, the labellum being uppermost.

Genus *Liparis*

Small green plants. Stem with a basal pseudobulb, leaves longer and narrower than in the preceding genus. Spurless flowers with an upward-pointing labellum in most cases, the perianth members being narrow and spreading.

Genus *Corallorhiza*

Saprophytic plants, with a knobbly root system resembling a piece of coral. The stem bears long sheathing scales at its base and is often greenish as it contains some chlorophyll-bearing cells. Flowers yellowish, the outer sepals curving downwards. Labellum short, white in colour with bright red spots.

Division B Basitonae

The stamen in plants in this division is attached by its base to the column occupying a position in front of the column. The viscidia lie on either side of the entrance to the spur, where it exists.

Genus *Herminium*

Small plants with a single sessile tuber and several stalked tubers. There are several narrowly ovoid basal leaves, and usually a single narrow stem leaf. The flower spike is dense, composed of numerous small yellow-green flowers. Floral segments narrow and connivent, the labellum being longer and three lobed, with a long central lobe. There is no spur.

Genus *Coeloglossum*

The plants have palmate tubers, oval base leaves, and narrower stem leaves. The flower spike is rather dense. The outer segments of the flower are united in a greenish hood, the labellum being strap shaped, with three terminal lobes, green in colour with a purple-brown tinge. The bracts are long and the spur is short.

Genus *Gymnadenia*

Plants with palmately lobed root tubers, stems bearing a number of long narrow leaves and long spikes of densely packed, sweetly scented flowers. The outer lateral perianth segments are spreading, the rest form a hood. The short labellum has three more or less equal lobes and bears at its base a very long slender spur. The two pollinia are borne on caudicles and attached to one long narrow viscidium.

Genus *Pseudorchis*

The fasciculated tubers are divided to the base. The basal leaves are flattened and oval, stem leaves lanceolate. The spike is dense with small yellowish-white flowers, the perianth members being more or less equal and connivent. The labellum is three lobed with a longer central lobe. The spur is short and blunt. The bracts are narrow and longer than the ovary.

Genus *Platanthera*

The tubers are entire and tapering. The basal leaves are broad oval in shape, the smaller upper stem leaves grading into the bracts. The flower spike is rather lax, the flowers white and strongly scented. The lateral outer perianth segments are spreading, the rest connivent into a hood. The labellum is narrow and strap shaped, the spur at its base being very long and narrow. The pollinia are borne on caudicles, and each is attached to a separate viscidium lying laterally to the entrance to the spur.

Genus *Neotinea*

The tubers are globular. The leaves are elongated and may bear rows of small dots, reddish purple in colour. The flower spike is very dense, the flowers facing in one direction. The perianth segments form a hood, the labellum being slightly longer and three lobed, the central lobe forked at its tip like a snake's tongue. The spur is short.

Genus *Ophrys*

The tubers are globular, the leaves elongated, blunt ended, and marked with parallel veins. The flower spike is open with a few large flowers. The perianth segments are spreading. The distinctive labellum, by which the genus is easily recognized, is large, entire or three lobed, often convex, velvety in texture, dark coloured with conspicuous markings. There is no spur. The column is long and erect, and the pollinia are borne on long caudicles.

Genus *Himantoglossum*

The tubers are broadly ovoid, the leaves numerous, oblong, and pale green. The whole plant is tall and massive, bearing many strong-smelling flowers. The perianth segments form a rounded helmet. The labellum is most curious and distinctive. It is three lobed, the lateral lobes being narrow and crinkled, while the very long central lobe, which is coiled like a watch-spring when in bud, unfurls and twists spirally as it does so. The spur is short.

Genus *Orchis*

The tubers are spherical or ovoid. The leaves are usually unspotted, the lower leaves forming a rosette, the upper leaves sheathing the stem. While the flower spike is emerging it is usually enclosed by thin spathelike leaves. The bracts are thin and membranous. All the perianth segments except the labellum are formed into a hood. The labellum is divided into three lobes, which may again be subdivided and it bears a spur.

Genus *Dactylorhiza*

This genus is separated from the genus *Orchis* by the palmately divided tubers. The basal leaves are sometimes spotted, do not form a rosette at the time of flowering, and the plants lack the thin spathe-like leaves over the emerging spike. The bracts are always leafy. The perianth segments never form a helmet as in the genus *Orchis*, but are either erect or spreading. The labellum bears a spur.

Genus *Aceras*

The tubers are ovoid, the lower leaves long, crowded, and unspotted, the stem leafy with the upper leaves grading into the bracts. The yellowish-green leaves are borne in a long narrow spike. The perianth segments apart from the labellum form a helmet, while the labellum is shaped like a man. The lateral lobes form the two long thin arms, and the central lobe is forked to form the two long legs. There is no spur.

Genus *Anacamptis*

The tubers are ovoid, the stem leafy, and the flowers borne in a conical spike. The lateral outer perianth segments are spreading. The labellum is three lobed, with two oblique erect plates acting as guides to the long slender spur. The two pollinia are borne on a narrow transverse viscidium, which curls round and grips the proboscis of any visiting insect.

Hybridization

Hybridization between orchid species is by no means an uncommon occurrence, being especially common among members of the genus *Dactylorhiza*. The resulting hybrids are usually difficult to identify and cause many a headache to the field botanist, but at the same time they are fascinating subjects to investigate.

The detailed study of the genetic mechanisms underlying the production of hybrids does not fall within the ambit of this book, but anyone wishing to make a deeper study would be well advised to read the excellent book by Dr C. A. Stace *Hybridization and the Flora of the British Isles*.

The following short chapter outlines some of the important factors involved, while a list of all the known orchid hybrids found in the British Isles follows the section on the individual species.

Before we can consider hybrids, it is first necessary to define a species. According to Dr Stace 'the species is a unit of practical value, visually recognisable and of evolutionary significance. Morphological and genetical data should be used in its recognition.'

A hybrid may be defined as the offspring of two different taxa. These may be either different species or different genera, in which case we refer to an interspecific or an intergeneric hybrid. No natural hybrids are known to involve more than two genera, and in the British Orchidaceae all known hybrids are confined to a single tribe.

Interspecific hybridization can lead to a new fertile species in one step but the influence of hybridization on evolution is usually far more subtle, and worked out over a long period. Interspecific hybridization is an essential step in evolution, providing a pool of genetic material from which new lines can emerge. Such lines will then develop or fail under the influence of natural selection. In hybrids characters can arise which do not appear in either parent.

Mayr (1942) describes two ways in which interspecific hybrids can arise. Firstly by 'primary intergradation' where two species are not sufficiently distinct to prevent hybridization, and secondly by 'secondary intergradation' where distinctness which once existed between species is now breaking down.

The first-generation hybrid (F1) may itself be diluted by repeated backcrossing with one of the parents, a process called 'introgression'.

The production of a fertile hybrid between two species should not be taken as evidence that they are closely related. It is not always possible to predict whether a hybrid will be fertile or infertile, and gradation of fertility does exist. Similarly, fertile hybrids between vastly differing taxa are known to exist, as witness some of the cultivated orchid hybrids.

In some cases plants may be physically and genetically capable of producing hybrids, but such hybrids will not occur in the wild because the species concerned are rare and widely separated geographically.

In describing hybrids it is customary to write the names of the parents in alphabetical order, or to quote the name of the female parent first—where it is known. Thus the hybrid between the Common Spotted-orchid and the Southern Marsh-orchid is written *Dactylorhiza fuchsii* × *praetermissa* while the hybrid between the Common Spotted-orchid and the Frog Orchid is written *Coeloglossum viride* × *Dactylorhiza fuchsii*.

Natural Hybridization

If you examine the orchid species which occur in Great Britain and the hybrids which have been recognized, then it becomes apparent that there are potentially a great number of hybrids still to be recorded. The formation of many of these potential hybrids is prevented by a number of isolating mechanisms or breeding barriers. Hybrids will not be formed if the ranges of the two species do not overlap or allow long-range pollen dispersal to occur. Man's activities may lead to the ecological isolation of a species because of the destruction of habitat.

Timing is also important. Hybrids cannot occur if the species concerned flower at different seasons, or even if their flowers open at different times of the day.

In the species of the genus *Ophrys* where pollination is effected by pseudocopulation the insects involved will only visit one particular species and are not attracted to other species of the same genus.

Hybrids cannot be formed between species where the pollinia and stigma are differently placed so that contact will not occur, and it will obviously be rare in those species which tend to be 'autogamous' (self-pollinating) or 'cleistogamous' (self-pollinating within a flower which does not fully open).

In some cases hybrids are not formed because of pollen incompatibility. The pollen tube may fail to grow or it may die, the egg may not be penetrated, or the gametes may fail to fuse. Even if fertilization occurs the hybrid may fail because of early embryonic death.

Where hybrids are formed there seems to be no correlation between fertility and vigour. Most F_1 orchid hybrids are fertile, but some, for example *Dactylorhiza fuchsii* × *D. maculata*, are highly sterile, while in others the degree of sterility can vary.

Hybrid vigour

Hybrid vigour is an example of the chance combination in an F_1 hybrid of the most favourable genetically controlled growth factors of both parents. Equally the combination of the parental genetic factors could be detrimental or even lethal—it is all a matter of chance. It also follows that in the F_2 and succeeding generations there will be a dilution of these characters. One splendid example of hybrid vigour comes to mind in a specimen of *Dactylorhiza fuchsii* × *D. incarnata* on Harlech golf-course. Fully three times larger than either parent, the massive pink spike stood up like a lighthouse in a sea of lesser orchids.

Hybrid swarms

A fertile F_1 hybrid may well give rise to an F_2 generation, but it can also backcross to either parent. If the F_1 is self-sterile ($F_1 \times F_1 =$ sterile) then it can only reproduce by such backcrossing or introgression, a procedure which will give rise to a range of plants intermediate in characteristics between the parents and the F_1 hybrid. This range of plant types is called a hybrid swarm. In nature this backcrossing usually occurs with one parent only, forming a population whose characteristics are clearly referable to that parent.

Environmental conditions may also dictate the composition of a hybrid swarm. One such group, of the parentage *Dactylorhiza praetermissa* × *D. traunsteineri*, showed a preponderance of *D. traunsteineri* types ten years ago. Since then the fen where they grow has progressively dried out and it is apparent that those plants with a growth habit more akin to *D. praetermissa* are at an advantage in the drier conditions. Natural selection has taken its toll, so that now the hybrid swarm is composed of plants primarily of a *D. praetermissa* type. Environmental conditions are rarely static and any viable plant population will show adaptation to changes in those conditions.

Identification of a hybrid

When assessing the possibility of a plant being a hybrid, one should bear in mind the three criteria established by Cockayne in 1923.
1. The plant should be morphologically intermediate between the putative parents.
2. It should grow in the field in proximity to both parents.
3. If it is fertile then there should be a segregation of types in the F2 and successive generations.

The difficulty of satisfying all these criteria is well demonstrated by Summerhayes (1968) in describing a hybrid between *Dactylorhiza fuchsii* and *Gymnadenia conopsea*. A bursicle is present in *Dactylorhiza* but absent in *Gymnadenia*. In the hybrid he found the bursicle to be present, imperfectly formed, or absent in different flowers on the same spike.

Description of Species

1 Lady's-slipper

Cypripedium calceolus L.

The Lady's-slipper has never been a common plant in Great Britain even in the past, since its beauty and strangeness attracted people to pick or uproot it. It was first recorded in 1640 at Helks Wood near Ingleborough, where it had all been dug up by 1796. Sowerby in his *English Botany* of 1873 remarks that it is 'very rare and now nearly if not quite extinct . . . occurred in several stations in Yorkshire . . . as lately as 1849'.

Old records refer to it growing in Cumberland and Westmorland in woods on limestone, although it was never frequent there. Most of the records are for the Magnesian limestone just north of Hartlepool in Durham, the area of Ingleborough in Lancashire, and most frequently on the Carboniferous limestone of the North and West Ridings of Yorkshire.

It is a large plant 30 to 50 cm tall, with up to five large, bright green, alternate leaves which sheath the stem. The leaves are strongly furrowed and bear prominent veins.

The bracts are large and leaf-like, standing up behind the flowers like erect hoods. Most stems bear only one flower, although plants with three have been recorded, the flowers being very large and distinctive. The sepals and petals are claret coloured and arranged in the form of a cross, the ventral arm of which is formed by the conjoined lateral sepals. The lateral petals are usually slightly twisted and are 3 to 5 cm long.

The labellum is very large, 2 to 3 cm long, shaped like a bright yellow slipper, the floor of the pouch being marked with lines of orange dots. The opening of the slipper is partly filled by the large spathulate staminode, which also bears brownish-orange dots. When the flower bud begins to open the slipper protrudes in a smooth yellow bulge between the longer, pointed, slightly hairy perianth segments. Flowering occurs in late May and early June.

The first leaves are formed in the fourth year after setting seed, but the mature plant is often sixteen years old before flowering first takes place. Fertilization is by small bees of the genus *Andrena*, which visit the orchid for the nectar secreted in the floor of the slipper. However, in the plants which still grow in the north of England, insect-pollination does not seem to take place, so that hand-pollination has been resorted to in the hope that fertile seed may be produced. The species has flowered recently in its known locality, while any other localities are likely to be the product of artificially produced plants of Continental origin.

The Lady's-slipper seems to favour steep grassy slopes below the limestone scars, where I have seen it growing in clearings between stunted oak, Hazel, and ash, in company with the delightful pink *Primula farinosa*.

No variation in form or colour has been recorded in this country although plants with white flowers have occurred in Switzerland.

Listed in the *Red Data Book*.*

2 White Helleborine

Cephalanthera damasonium (Miller) Druce

[*C. latifolia* Janchen]

The White Helleborine is by far the commonest of the three members of this genus, and is often to be found in large numbers in beechwoods on chalk in the Home Counties. It makes a fine showing on the bare floor of the woodland, the creamy-white flowers contrasting with the rich brown of the dead beech leaves.

The flowering stem ranges from 15 to 60 cm in height, arising from a mass of fibrous roots which penetrate vertically downwards for as much as 40 cm. The taller plants are found on the edge of clearings where there is more light.

The oval leaves are borne in two rows up the stem, merging into the leaf-like bracts which are longer than the ovary. The leaves are strongly grooved.

The spike carries three to sixteen ivory-white flowers set vertically close to the stem. The sepals and petals never diverge to any extent, giving the impression that the flowers do not open properly. The labellum is hinged in the middle, the hypochile being yellow within, while the epichile is heart shaped with five dark yellow ridges on its upper surface. The ridges seem attractive to visiting insects, a fact first noted by Darwin, and they often appear tattered where little pieces have been nibbled away.

There is no rostellum, a distinguishing feature from the genus *Epipactis*, and the anther is suspended from the tip of the long column. Most flowers are self-pollinated, although a few are pollinated by bees. In either case it seems highly efficient, since most plants carry at least three ripe seed capsules. In this species the ovary is deeply grooved, broadest just short of its distal end, and untwisted, which differentiates it from the narrow twisted ovary of *Cephalanthera longifolia*.

After seed is set as many as eight years may elapse before an aerial stem arises, and a further two to three years before the plant flowers. In any large colony quite a number of immature, non-flowering stems can be seen.

The White Helleborine is relatively common in the south-east of England, especially in woods along the North Downs and in the Chilterns. It is absent from Wales, Scotland, and Ireland, and it is not found west of Somerset, apart from one locality in south Devon. It thrives in bare beechwoods, often in company with Fly and Bird's-nest Orchids, and will even be found in hedgerows and near isolated beech trees.

It flowers from the last week in May until the end of June, although in one exceptional spring I found it fully out on 30 April.

Variants with double lips and columns are not infrequent while, in 1924 at Horsley, Surrey, C. B. Tahourdin recorded a plant with white stem and leaves.

The hybrid (*Cephalanthera* × *schulzei* Camus, Berg & Camus) between this species and the Narrow-leaved Helleborine is known in France and Germany and was first recorded in this country in May 1974, when John Lansley and I found two plants in flower in a wood in Hampshire.

3 Narrow-leaved Helleborine

Cephalanthera longifolia (L.) Fritsch

The Narrow-leaved Helleborine with its open spike of pure-white flowers set above the long narrow leaves is, without doubt, a far more elegant plant than the White Helleborine.

It is much the same size, 15 to 60 cm in height, with a mass of long deep roots, some rather fleshy, the others being small and wiry. The sword-shaped leaves are pale green and more numerous than the leaves of *C. damasonium*. The bracts are smaller than the ovary, which is another distinguishing point.

The stem bears up to twenty white flowers in a long open spike, the lower flowers being borne nearly horizontally and well away from the stem. The perianth members are relatively longer and more spreading, so that the labellum is more exposed. The grooves on the epichile are three in number and dark orange. The plants seem rather sensitive to frost and a late cold snap will often result in flowers disfigured with brown spots, these flowers falling off at the slightest touch.

The ovary is slim and cylindrical, the grooves being twisted through 180° anticlockwise.

The flowers are insect-pollinated, chiefly by small bees, and self-pollination occurs but rarely since the pollen mass cannot fall forward on to the stigma. Seed production in this species is rather poor and few ripe capsules will be found on any one plant.

The Narrow-leaved Helleborine is much rarer than the White Helleborine and has a curiously patchy distribution. Although it existed before 1930 in several sites in Kent, Sussex, and in the north of England, it is now recorded chiefly from Hampshire and the Chilterns in the south, parts of Wales, Cumberland, and western Scotland from Kintyre to Sutherland. It is thinly scattered throughout Ireland.

Within the last ten years new sites have been found in North Wales and in Scotland as far north as Wester Ross and with further searches we may find more of this delightful helleborine. One of the Irish sites in Mayo is most curious—the plants, which are of very small stature, grow in blown calcareous sand over peat in a rocky area near the sea.

However, most of the sites are in beechwood on chalk or mixed ash and oak on limestone, the woods being more open and light than those favoured by the White Helleborine. Consequently the Narrow-leaved Helleborine is found growing in competition with a host of other plants, often under brambles and scrub. The flowering period, throughout May, is slightly earlier than for the White Helleborine.

Variations are seldom recorded and the hybrid between *C. longifolia* and *C. damasonium* has been noted under the latter species.

4 Red Helleborine
Cephalanthera rubra (L.) Richard

Many of our orchids are strange or brightly coloured, but none can rival the Red Helleborine, combining as it does a graceful growth habit with a pink colour of extraordinary intensity.

It has always been a rare plant in this country, but it is so unmistakable that it is not necessarily wise to discount the unconfirmed reports of its past whereabouts.

The stem is from 20 to 60 cm high, plants in this country normally being smaller than those on the Continent. The root system is composed of thin horizontal roots, unlike those of the two other *Cephalanthera* species, and from this root system buds arise to give aerial shoots, which may in turn separate to form satellite plants.

The leaves are fairly short, limp, and dark green, while the slender bracts are longer than the ovaries. Up to fifteen flowers have been recorded on one stem, although the average is probably nearer five, the spike of rather large buds and flowers having a slight resemblance to a freesia.

The perianth segments are long, pointed, and a fine pink with just a shade of lilac. The labellum is pink with a paler, yellowish-white centre, the epichile bearing five to seven orange ridges.

The ovary is narrow, cylindrical, and twisted, and like all the floral parts is covered with fine glandular hairs. The flowers are insect-pollinated and M. J. Godfery records several species of bees visiting plants on the Continent. I have seen flowers in England visited by a number of small hoverflies and once by a Small Skipper (*Thymelicus sylvestris*).

The Red Helleborine favours beechwoods on chalk or limestone and seems to thrive even close up to the trunks of mature trees. However, its capacity to flower is evidently dependent on incident light, and in many cases the plants now fail to flower. In the vegetative state it is capable of surviving for many years, coming into flower if conditions of cover or shade alter to its liking. Field-work is currently being undertaken on the Continent to try to discover the optimum conditions for flowering to take place. The flowering period is from late June to the end of July.

Old records include Kent, Sussex, Somerset, and, even very dubiously, Gairloch in Wester Ross. C. B. Tahourdin in 1924 wrote that the Sussex station had probably been destroyed, but it is not beyond the bounds of possibility that the felling of woodland may lead to its re-emergence in one of its old haunts.

The two main centres of distribution now are in the Chilterns and in Gloucestershire. The Chilterns site was discovered in 1955 and in 1956 R. S. R. Fitter found ten plants in flower with sixty-four blind plants. Five flowering spikes were produced in 1986 after the felling of some trees. In one of the Gloucestershire sites it flowers fairly regularly, but in the other sites it has not flowered recently. It was rediscovered in Hampshire in 1986, and two spikes were produced in 1987.

No varieties or hybrids have been recorded in Great Britain although albinos are known in Europe.

Listed in the *Red Data Book*.*

5 Marsh Helleborine

Epipactis palustris (L.) Crantz

The Marsh Helleborine probably comes as close as any other orchid species to the popular conception of an orchid flower, since it resembles a miniature *Cymbidium*.

It is a plant of moderate stature 20 to 60 cm tall, arising from an extensive system of shallow creeping rhizomes. The leaves are numerous, lanceolate in shape, and folded. They have three to five prominent veins and the bases of the leaves and stems often bear violet sheaths.

The spike is lax and one sided with as many as twenty rather large flowers, the lower bracts exceeding, and the upper bracts being shorter than, the ovary. The ovary is narrowly pear shaped and untwisted, although its stalk is twisted through 180° so that the labellum of the flower lies ventrally.

The outer perianth segments are pointed and purplish brown. The upper inner perianth segments are white, tinged with pink near the base.

The labellum is white and divided into a hypochile and epichile, the two being joined by a hinge. The cup of the hypochile is marked with parallel red veins and has two ear-shaped lateral lobes. The epichile is broad and white with a frill around the edge and an erect yellow plate across the base.

This yellow plate secretes nectar into the cup of the hypochile, attracting ants, bumble-bees, and hoverflies. Recent work by L. A. Nilsson on the island of Öland in southern Sweden has shown that solitary wasps are the most important pollinators, although their visits are relatively infrequent. He suggests that the flowers are particularly adapted to pollination by male solitary wasps of the genus *Eumenes*, and that while trying to balance on the epichile the head of the insect touches the sticky viscidial pouch, which ruptures and glues the pollen mass on to the insect's head. There was no evidence that the hinged labellum actively catapulted the wasp against the rostellum, as had been suggested by Darwin. When the wasp visits the next flower the pollen masses adhering to the head contact the stigma lying under the rostellum. Ants also are important pollinators.

This method of pollination is efficient and more frequent than self-pollination. Despite this, it would appear that within the orchid colony most multiplication is vegetative, by development of new aerial shoots from the branching rhizome, seed production contributing primarily to the establishment of fresh colonies.

The Marsh Helleborine is widely distributed throughout England, Wales, and Ireland, and less frequently in southern Scotland, but from a study of old records it is evident that the number of colonies is steadily decreasing as its wetland habitat is drained. It prefers wet areas where there is dissolved lime in the water and it is most frequently found in dune slacks where the shell sand provides the necessary lime. For this reason it is to be found at its best near the coast of east and south England and in west Wales. There are two most interesting records, one in Wiltshire and one in Bedfordshire, where the Marsh Helleborine grows on chalk downland (per T. C. E. Wells).

Although the flowering period is from early July to early September, it is usually to be found in full flower in the last week of July.

No hybrids have been recorded but there are two well-known variants. A dwarf form flowers in a chalk-pit on the south side of the Thames—the flowers are of the normal size and colour, although the plants themselves are barely 10 cm high. Unfortunately, the numbers of flowering plants have decreased markedly over the last twenty years as the birch trees and dogwood have grown up. I understand that plants removed some years ago from this site maintained their dwarf form when cultivated elsewhere.

The other form, var. *ochroleuca*, lacks the red-brown pigments in the flower so that the outer perianth segments are greenish yellow, the inner segments and labellum remaining white. I have seen it in South Wales flowering with the normal form and it has also been recorded from Devon, Cheshire, and Ballyvaughan, Co. Clare, in Ireland.

6 Broad-leaved Helleborine

Epipactis helleborine (L.) Crantz

[*Epipactis latifolia* (L.) Allioni]

The Broad-leaved Helleborine is a large strong-growing plant, the stem arising from a thick cluster of relatively short roots. Plants are frequently to be seen up to 90 cm in height, occasionally several stems arising close together, but never in the showy clumps of *E. purpurata*.

The leaves are broad oval in shape, dark green, strongly ribbed and furrowed, and arranged spirally up the base of the stem. They are well grown by the end of May. The lower bracts are the same length as the flowers, while the upper bracts are shorter.

The flower spike tends to be one-sided and can be quite imposing with up to a hundred flowers, but it is a very variable plant, and size will depend to a great extent on the flowering site.

The sepals are green, rather broad and blunt, and overlap at the base so that the whole flower is cup shaped. The upper petals are slightly shorter and pink, especially at their tips.

The hypochile of the labellum is cup shaped, dark red brown within, and lined with sticky glistening nectar. The epichile is pinkish brown, especially near its apex, and the tip is reflexed. Near the base of the epichile are two bosses which are rough and brown. There has been much dispute over the years as to whether this feature can help in identifying the species and I think it can reasonably be accepted that it does. Despite the acknowledged variability of the flowers of *E. helleborine*, a careful study of many hundreds of specimens has shown it to be a feature of more than 80 per cent of the flowers. By contrast, in the Violet Helleborine nearly all the flowers have smooth pink bosses. It is essential that this examination be made on freshly opened flowers, before any of the floral structures start to wither.

The rostellum is primitive, large, white, and persistent. The ovary is smooth and hairless.

Most of the problems of identification arise because of the variability of the species, its size and colour depending on the flowering site. In deep shade plants tend to be less robust with paler, greener flowers, while in dry exposed places the plant is more dwarf and the flowers more heavily flushed with pink. However, the characteristics of the other *Epipactis* species are sufficiently distinct for there to be no doubt in one's mind once the flowers of *E. helleborine* have been seen and closely examined. The well-tried maxim of assuming an unknown plant to be the commoner rather than the rarer species certainly applies in this case.

The Broad-leaved Helleborine is pollinated by wasps and takes at least eight years to reach maturity from seed. It is widely distributed throughout the British Isles, extending to the far north of Scotland, but is less frequent in East Anglia, Scotland, and Ireland.

In the south of England the Broad-leaved Helleborine is to be found flourishing in beechwoods, especially on the verges of roads through them, but in the north and west it can be found on limestone pavement and on rocky screes with *E. atrorubens*. The flowering period is from mid-July to early September, about a fortnight earlier than the Violet Helleborine.

Hybrids have been recorded in this country between this species and *E. purpurata*, *E. atrorubens*, and possibly *E. leptochila*.

7 Violet Helleborine

Epipactis purpurata Smith

[*E. sessilifolia* Peterman; *E. violacea* (Dur. Duq.) Bor.]

Although the Violet Helleborine is less widely distributed than the Broad-leaved Helleborine, its habit of growing in clumps of as many as twenty spikes makes it a most attractive plant. The rhizomes are vertical and deep rooting, in a cluster of as many as fifty.

The size range is the same as that of *E. helleborine* but it is usual for plants to be smaller and slimmer even when they grow in the same locality. The leaves are much narrower and distinctly grey green. The sheathing bases of the leaves and the stem are often suffused with a fine violet colour, and in some plants this is so extensive that the leaves are entirely streaked and coloured violet—the so-called Violet-washed Helleborine.

The flowers are in a one-sided tightly packed raceme, the narrow bracts equalling or exceeding the flowers, thus giving the flowering plant a very leafy appearance. The sepals are green and overlap at their bases, but they are more pointed and spreading than those of *E. helleborine*. The upper petals are of a much paler whitish green.

The labellum is of a similar shape to that of *E. helleborine*, but the lining of the hypochile is a much paler pinkish brown. The epichile is pale pink, especially towards its tip, which is recurved, and the two bosses at its base are bright pink and smooth in texture.

The rostellum is white and persistent, while the ovary in this species is covered in short hairs and feels rough to the touch. Pollination is effected by wasps.

The distribution of the Violet Helleborine is restricted to south-east and south England, the greatest density of population being in parts of Kent, Surrey, and the Chilterns. It is a plant chiefly of beechwoods on chalk, more rarely of oakwoods on a sandy soil, and flourishes in very much darker places than the Broad-leaved Helleborine can tolerate.

I have found it growing in abundance at the bottom of an abandoned chalk-pit, where the scrub cover was so dense that little else would grow under it.

It is also remarkably persistent and in one Surrey colony robust plants pushed up through a newly laid tarmac pathway and even flowered between the concrete supports of a fence.

It flowers two weeks or so later than *E. helleborine* in August and September, but the flowering periods overlap for about a month. Hybridization with *E. helleborine* has been recorded.

8 Slender-lipped Helleborine

Epipactis leptochila (Godfery) Godfery

[incl. *E. cleistogama* C. Thomas]

The Slender-lipped Helleborine is a fairly recent addition to our flora, the first description having been published by M. J. Godfery in 1919 in the *Journal of Botany*. For some years after that it was thought to be restricted to Great Britain, but it would now appear to be moderately widespread in western and central Germany and in Denmark. Further field-work will doubtless result in an extension of its recorded range.

The stem, which is from 15 to 60 cm in height, arises from a deep, thick root-stock, and in mature plants as many as six stems may be formed on one plant.

The broad yellow-green leaves are carried in two rows—a distinguishing feature from *E. helleborine*—and are rather floppy. The loose flower spike may carry as many as twenty-five pendulous flowers set all round the stem, the lower bracts longer and the upper bracts shorter than the flowers.

The flowers are relatively large and all the perianth segments are long, pointed, and spreading, the sepals scarcely overlapping at their bases. Apart from the labellum, all the parts are green and lack the red or purple colouring which may be present in the previous two species.

The labellum is green with a faint pink flush in some cases. The hypochile is pale brown within. The epichile is extremely long and sharply pointed, the tip not being reflexed, and from this the plant gets its name. At the base of the epichile are two rather large green-coloured bosses. The flowers lack a rostellum, the anthers being borne on a stalk (clinandrium) in such a way that, when viewed sideways, there is a distinctive hole visible between the column and the anther. The only time confusion may arise is when one finds an etiolated, spindly specimen of *E. helleborine* devoid of reddish coloration. I have found such plants growing under heavy shade, and the tip of the epichile is not always reflexed. However, the pointed epichile of *E. leptochila* is fully half as long again, and there should be no doubt when considering this feature in conjunction with the other structural differences between the species.

E. leptochila is usually autogamous, although A. J. Richards observed wasps (*Vespa germanica*) in Northumberland removing all or most of the pollinia. The Slender-lipped Helleborine is sparsely distributed, most often in beechwoods on chalk, in east Kent, Surrey, Hampshire, the Chilterns, Gloucestershire, and Devon. *E. leptochila* and *E. dunensis* are closely related and usually autogamous. They were thought to be ecologically, distributionally, and morphologically distinct. Since 1974 A. J. Richards and G. A. Swan have found inland populations, mainly on lead- and zinc-polluted woodland sites in Lincolnshire, Yorkshire, and Northumberland, which exhibit considerable grading of the characters of both *E. leptochila* and *E. dunensis*. *E. leptochila* is often found accompanied by *E. helleborine* and *E. purpurata*, but seems to favour more shaded places. In one of its Kentish sites it is flourishing under a dense covering of Hornbeam.

The flowering period is earlier than that of *E. helleborine*, being from the very end of June throughout July.

A possible hybrid with *E. helleborine* has been recorded.

9 Dune Helleborine

Epipactis dunensis (T. & T. A. Stephenson) Godfery

The Dune Helleborine is one of our rarest helleborines. It has long been known from dune sites in Anglesey, Lancashire, and Northumberland. Since 1974 a number of inland populations have been found in Northumberland and Lincolnshire, where it grows with *E. leptochila* and *E. helleborine*.

In stature it is much the same size as *E. helleborine*, plants in sheltered places near trees tending to be larger than those growing in exposed dunes. The root-stock is woody and deep rooted, bearing a few wiry rhizomes. The base of the stem has a loose funnel-shaped sheath around it, unlike *E. helleborine*.

The leaves are yellowish green, broad, and stiff, and are borne in two rows up the stem. The bracts are shorter than the flowers.

The open flowering spike bears up to twenty flowers, the individual flowers standing out horizontally at first, although they tend to droop as the seed capsules ripen. The flowers are rather small and dingy, the perianth segments being short and blunt, so that the flowers appear to be cup shaped and not to open properly. All the perianth segments are greenish yellow, while the labellum is pinkish white with a green tip. The hypochile is darker pink, the epichile being rather broad with a recurved tip and two low bosses at its base

The rostellum is often absent or withers early, so that the flowers are self-fertilized.

The Dune Helleborine grows on the Continent in coastal dunes in north-west France, Belgium, and Germany. In Britain it grows both in classic wind-swept dunelands and inland woods, on river-gravels and by mine-workings, where the soil is heavily polluted with lead and zinc. The populations of these inland sites have characters of both *E. leptochila* and *E. dunensis*. The lead and zinc levels are so high that they kill much of the stronger vegetation. The helleborines—protected by their mycorrhizal fungi from the full toxic effects—thrive on the lack of competition.

The flowering period is from late June to the end of July; the best time to see the Dune Helleborine is the second week of July. Like so many other dune-dwelling plants it is affected by drought, and in a very dry summer the flowers quickly shrivel and discolour.

No varieties or hybrids have been recorded in this country.

Listed in the *Red Data Book*.

9a Young's Helleborine

Epipactis youngiana A. J. Richards & A. F. Porter *sp. nova*

See Postscript, pp. 140–1

10 Pendulous-flowered Helleborine

Epipactis phyllanthes G. E. Smith

[incl. *E. vectensis* (T. & T. A. Stephenson) Brooke and Rose; *E. pendula* (C. Thomas); *E. cambrensis* (C. Thomas)]

The Pendulous-flowered Helleborine was first described by G. E. Smith in the *Gardeners' Chronicle* of 1852. Not until a century later was it shown by D. P. Young that the other three plants listed above were, in fact, all forms of one good species, *Epipactis phyllanthes*.

The Pendulous-flowered Helleborine is rather a small and delicate plant in comparison with others of the same genus, having a height of 10 to 40 cm. Most plants have a single aerial stem, but in North Wales plants with up to three stems have been described.

The base of the stem is sheathed, the uppermost sheath being open mouthed and funnel shaped. In woodland plants in Kent the base of the stem is frequently dark violet.

As many as sixteen leaves are borne in two opposite ranks, the largest leaves being above the middle of the stem; the leaves are rounded, stiff, and ribbed. The bracts are longer than the flowers.

The stem bears up to twenty drooping flowers in a dense spike. The flowers scarcely open and indeed many of the upper flowers of the spike remain as tight buds and never open at all. This is especially noticeable in the duneland plants, where the dry conditions result in a scorched and withered spike long before the upper flowers have had a chance to open.

The outer perianth segments are green and pointed, much longer than the equally pointed upper petals. The labellum is variable, but tends to be rather short with a reflexed tip. Occasionally it may be tinged pink or lilac. It is, however, scarcely visible, since the outer perianth segments cluster round it.

The flowers are always self-fertilized, and in many cases are cleisto-gamous—fertilization occurring within the unopened flower. The Pendulous-flowered Helleborine has now been widely recorded in south England, in North and South Wales, and in scattered localities in north England. Fifty plants were found in 1974 by A. J. Richards and G. A. Swan, growing with *Epipactis helleborine* under birch trees in heavy shade, by the spoil heaps of an old lead mine in the area of the South Tyne, this being the most northerly record for England. It is also known from a handful of localities in north and south-east Ireland.

Comparison between the woodland plants of north Kent and the duneland form from North Wales is most interesting. The north Kent plants grow on chalk under a mixed copse of beech, chestnut, and larch, the flowering spikes emerging through a thick carpet of Ivy and Dog's Mercury. The plants are small and slender, scarcely 20 cm high, with two to three dark-green leaves and up to thirteen flowers. They flower in late July and August. The colony had decreased from 121 in 1957 to 10 in 1975 as the woodland became overgrown.

The Welsh sand-dune plants by contrast are yellower and more robust, with stems up to 40 cm and as many as six leaves. The leaves are bigger and yellower, and far more coarsely ribbed. The spikes bear as many as nineteen flowers which are identical with those of the Kentish plants, and similarly the upper buds have the tendency to fail to open. The flowering period is earlier, from late June onwards. The plants grow on low dry hummocks among Creeping Willow, Bird's-foot Trefoil (*Lotus corniculatus*), and Red Fescue (*Festuca rubra*) and show no tendency to invade the wetter dune slacks or to grow in the dry sand. Unlike the Kentish plants, the colony is thriving, having increased from 144 plants in 1959, when P. M. Benoit first found them, to 318 in 1972.

No hybrids with *E. phyllanthes* have been reported.

11 Dark-red Helleborine

Epipactis atrorubens (Hoffman) Schultes

[*E. atropurpurea* Rafinesque; *E. rubiginosa* Crantz]

The Dark-red Helleborine can almost be identified by its habitat alone, being found on bare limestone cliffs and pavements, the only other helleborine occasionally found in such localities being *E. helleborine*. Although dwarfed in exposed sites, it varies from 15 to 60 cm high, with a stiff wiry stem arising from a short, thick root-stock. The roots are numerous, long, and thin, admirably adapted to penetrate the fissures of the limestone rock in search of water and nutrients.

The leaves are dark green, markedly folded, and are borne in two rows, the lower leaves being oval and the upper ones elongated and rough to the touch. The bases of the bottom leaves may be suffused with red.

The stem is densely covered in downy hairs, and the bracts equal the flowers.

The flowering stem may carry as many as twenty flowers in a sheltered locality, but usually there are fewer than ten to a spike. The flowers of different populations vary greatly. Some are small and cup-shaped, while others have long, elegant perianth segments. The caruncles on the base of the broad, recurved labellum are highly variable. All the floral parts are a deep wine red, unlike the colour of any other helleborine except the occasional dark-coloured specimen of *E. helleborine*, which differs markedly in its structure.

The ovary is distinctly downy.

The Dark-red Helleborine is to be found on limestone cliffs and pavements in west Yorkshire, the southern Lake District, North Wales, Wester Ross, Sutherland, Skye, and Banffshire. It can be seen to advantage in the Burren area of Galway and Clare, where it grows with the Dense-flowered Orchid. Its brilliant colour looks superb against the stark whiteness of the limestone rocks, where it often flowers on virtually inaccessible ledges or in deep fissures in the cliffs.

Pollination is by wasps, bees, and hoverflies, and I have seen plants in the Pennines at 570 metres visited by large hoverflies. The flowering period varies according to the locality, most flowers being fully out in mid-July.

Variation within the species is not great, plants usually being dark red in colour, although occasional plants in very exposed sites may have rather a washed-out appearance, with a tinge of green. In Wester Ross A. A. Slack and A. McG. Stirling have recorded plants with cream-coloured flowers. In the same locality many of the normally coloured plants have flowers with pointed sepals much longer than the petals, giving the flowers a more open appearance.

The hybrid between *E. atrorubens* and *E. helleborine* has been recorded in Denbighshire, West Yorkshire, and West Sutherland where the two species occur together. Since both are pollinated by wasps and have similar floral structures, this is not surprising, but the identification of such hybrids is difficult.

12 Ghost Orchid

Epipogium aphyllum Swartz

[*E. gmelini* Richard]

The Ghost Orchid has always been a plant of extreme rarity in this country which, combined with its erratic flowering habits, has often led botanists to think that it had become extinct. Even now it must be considered a great good fortune to be privileged to see this extraordinary orchid in flower.

The first reported flowering was in 1854 at Tedstone Delamere on the border of Herefordshire and Worcestershire. Unfortunately, the plant was dug up and transplanted to a garden where it promptly died and no others have ever been seen in the original site.

It was next reported in 1876 near Ludlow in Shropshire, flowering again in 1878 and 1892, while a further colony was found near Ross-on-Wye in 1910. On 19 September 1982 a solitary plant bearing two flowers was found in Herefordshire by Dr V. Coombs.

All other recent sightings have been in the Chilterns, where it was seen in 1924 by Mrs V. N. Paul, then a schoolgirl. Another plant was found in 1926, and then in 1931 a large spike was found in another wood in the same district. This plant was growing out of the middle of a decayed tree stump and Mrs Paul noted that it was 23 cm tall and bore three flowers. A second smaller plant was found nearby in 1933.

After that no plants were seen in flower until 1953. To quote Mrs Paul's report: 'Mr R. A. Graham had been searching for *Epipogium* for twenty years when in 1953 he was in a Buckinghamshire wood about ten miles from the Oxfordshire site. He was lighting his pipe when, over the bowl of it, he saw a plant of Epipogium growing among the beech leaves.' History does not relate whether he actually got his pipe to light!

There were in fact twenty-four flowering plants in the wood—until then only eleven plants had ever been seen in flower in Great Britain. A careful recheck of the 1931 site revealed two small plants in flower. Since then the Ghost Orchid has been found in flower in most years in one site or the other. In 1978 and 1979 several plants were stolen.

The underground system consists of a curiously flattened and branched rhizome which resembles a mass of coral. There are no roots but long slender runners arise from the rhizome mass and penetrate some distance from the parent plant. These runners bear scale-like leaves at intervals, and from these new buds form and develop into fresh plants. The runners soon die away and new ones are produced from the parent plant each year. The extent of the underground rhizome system is immense, and I have found them penetrating a rotten tree stump at least 100 metres from the nearest recorded flowering plant.

The stem is from 5 to 25 cm in height, with a markedly swollen base. It is pinkish in colour, marked with darker spots and broken lines, and usually ends in a blunt process above the uppermost flower. The bracts are short and membranous.

The flowers are large, up to six on a stem, on short untwisted stalks. The ovary is globular and similarly untwisted, so that the labellum usually points upwards.

The sepals and petals are narrow, yellow with minute red spots, and tend to hang drooping together. The labellum is divided into three lobes, the centre lobe being large and triangular with a wavy margin. It is pink with purple streaks and ridges and bears a fat swollen spur, likewise streaked with purple.

The spur wall contains nectar which attracts humble-bees. The rostellum is very sensitive and 'explodes' when the bee brushes against it, gluing the pollinia on to the insect's body. Cross-fertilization occurs when the bee visits another flower but seed is rarely produced in this country. The lowest of the four flowers of the plant which I photographed in 1970 did, in fact, have a ripe seed capsule.

The Ghost Orchid is restricted to well-established beechwoods on chalk, where it can grow in the deep humus formed by the rotted leaves. The colour of the flowers is such that they are extraordinarily difficult to see in the dim light against the background of dead beech leaves. Ingenious botanists have been known to locate the flowering plants by going out at night with a powerful torch, and shining it parallel to the ground in order to spotlight the spikes above the beech leaves.

The flowers are said to have a scent reminiscent of bananas but I fear that it is more like the smell of new gym-shoes.

The flowering period is extremely variable, ranging from June to October, and is said only to follow a thoroughly wet spring. The plants at the 1931 site tend to flower in September, rather later than those at the other site. The Ghost Orchid suffers badly from attacks by slugs, which frequently chew through the stems unless they are protected.

No variants or hybrids are known.

Listed in the *Red Data Book*.*

13 Autumn Lady's-tresses

Spiranthes spiralis (L.) Chevallier

[*Spiranthes autumnalis* Richard]

Records of the Autumn Lady's-tresses appear as far back as the sixteenth century and although it is not so widely distributed now as formerly, this charming but diminutive orchid can still be found in large numbers in a good year.

The one or two tubers may be ovoid or parsnip shaped, one tuber which has supplied the current growing plant being shrivelled, while the other is fat and white and will supply growth for the next season. New tubers are formed at the base of the stem as small finger-like projections, which enlarge and grow downwards. There are no roots, but it appears that the tubers carry a mycorrhizal infection.

Four to five bluish-green leaves are borne in a flat rosette, the leaves forming in early September, and persisting over winter to the early part of the following summer. The rosette then withers and the flower spike emerges from its centre. By then the new rosette has formed beside it and will, in due course, supply the next flower spike so that the flowering spike and the leaves appear to be unconnected.

The flat rosette of leaves is well adapted to the short turf in which it grows, proving remarkably weather resistant and little damaged by grazing. Treading by cattle does damage plants, while rabbits and sheep will eat off the flower spikes.

The stem is short, from 5 to 15 cm, is densely covered in white hairs, and bears several pale-green, sheathing, bract-like scales. The bracts equal the ovaries and sheath them.

The flowering stem bears up to twenty flowers arranged in a single tight spiral. The small flowers are trumpet shaped, the upper half of the trumpet being composed of the coherent white sepals and petals, and the lower half by the gutter-shaped labellum. The labellum is white edged and frilled, the central part being greenish and glistening with nectar. There is no spur.

The flowers have a faint honey scent and attract humble-bees. Cross-pollination is ensured by the position of the rostellum, which only hinges up to expose the stigma some time after the pollinia have been removed.

Seed production is efficient, and plants also multiply vegetatively by the production of lateral buds, but the process is slow and accounts for only a small percentage of the population.

It has been said that development is slow from seed, the first tuber being formed in eight years, the first leaves in eleven years, and the mature plant flowering in thirteen to fourteen years. However, T. C. E. Wells, in an intensive study of the species, has found that under laboratory conditions plants in culture media produced a leaf in the first year after germination, and flowered in five years.

The Autumn Lady's-tresses is chiefly to be found in old chalk pastures or in calcareous grassland where the grass is short. It frequently appears in large numbers on lawns, and seems equally at home in the fine turf of clifftops by the sea.

The main area of distribution is the southern half of England and Wales, especially south-east and south England, but it is also to be found scattered throughout most of the southern part of Ireland and the limestone area of the Burren. In one interesting area of north Norfolk it used to flower in grassland within a few metres of the Creeping Lady's-tresses growing in a pinewood—a unique conjunction.

The numbers may appear to fluctuate considerably from year to year, flower spikes emerging in thousands on a lawn or pasture, only to disappear again for many years. However, studies have shown that this is not truly the case, the total population remaining more or less constant, although it is difficult to realize this since the non-flowering rosettes are very hard to detect in the turf. After flowering the plant may even pass one or more years underground with no aerial structure, and then subsequently emerge to flower again. During that period it is thought that the tubers are re-forming with the assistance of the mycorrhizal fungus, so that if a plant is not visible above ground one cannot assume that it is dead.

The factors which cause the plants to flower are not understood, but weather, length of grass, and mycorrhizal activity in an otherwise rootless plant must all play a part.

The flowering period is from mid-August to the end of September.

There are no varieties or hybrids.

14 Summer Lady's-tresses

Spiranthes aestivalis (Poiret) Richard

First recorded in 1840, the Summer Lady's-tresses has always been restricted to two small areas of the New Forest. A plant found in 1854 in a bog in the Forest of Wyre in Worcestershire was described at the time as *Spiranthes aestivalis* but recent examination of the preserved specimen showed it to be an abnormal plant of *Gymnadenia conopsea*.

In 1900 nearly 200 plants were found in flower, but by the early 1930s the number had fallen to 20. Part of the cause of this decline has probably been drainage of the habitat, but plants were certainly dug up. In 1924 C. B. Tahourdin recorded that he had been sent a plant from the New Forest and commented, 'I am told it is nearly extinct there now'.

The last photograph appears to be that taken by A. B. Dalley of three plants flowering in July 1937, and there are unconfirmed records up to 1952.

While I was in practice in the New Forest I met the wife of an eminent Lepidopterist, who had an interesting story to tell. In late July 1959 she was out with her husband in the Forest and found a *Spiranthes* in full flower in a wet tussocky field, the area being within 800 metres downstream from one of the classic sites. *Spiranthes spiralis* does not grow in that area, and the habitat was quite unsuitable for it. No identification was made, as it was assumed to be the Autumn Lady's-tresses, but the possibility exists that *Spiranthes aestivalis* may still linger on near its old haunts.

The roots are fleshy and tapering, like those of *Spiranthes spiralis*, but the linear basal leaves cluster round the base of the stem. The hairless stem is 10 to 40 cm high, and carries true stem leaves which are long, narrow, and yellow green. Higher up the stem these grade into the leaf-like bracts, which are longer than the ovary.

The flowers, five to twenty in number, are carried in a single spiral row which is not as tightly twisted as the spike of *Spiranthes spiralis*. The flowers are similar in colour, but are longer and narrower. The labellum is oval with a wavy edge, and the ovary is smooth and hairless. The flowers are said to be faintly scented towards evening and the assumption has been made that they are pollinated by moths, although it has never been proved.

The Summer Lady's-tresses grows in marshy or boggy ground near springs or streams, with Purple Moor-grass (*Molinia caerulea*), sedges (*Carex* spp.), *Sphagnum* moss, and small amounts of Bog Rush (*Schoenus*). The soil is either neutral or only slightly acid, and the habitat is definitely wet.

The flowering period is from mid-July to mid-August.

Listed in the *Red Data Book*.

15 Irish Lady's-tresses

Spiranthes romanzoffiana Chamisso

[incl. *Spiranthes gemmipara* (Smith) Lindley]

The Irish Lady's-tresses has one of the most interesting distributions of any of the British orchids. In Europe it is restricted to the British Isles and more properly belongs to the flora of Canada and North America. In some of the areas where it occurs the Blue-eyed Grass (*Sisyrinchium bermudiana*) and Pipewort (*Eriocaulon aquaticum*) have been found, close relatives of plants which also belong to the North American flora. This has led to the theory that the Irish Lady's-tresses is a relic of the flora which was continuous across this whole area prior to the last Great Ice Age which ended 10,000 years ago. However, there is some doubt that the Pipewort of north-west Scotland is identical with that of North America.

In recent years the Irish Lady's-tresses has been discovered in a number of new localities in Great Britain, which markedly extend its known range.

It was first reported in 1810 in Ireland in West Cork and South Kerry, and a second large area of distribution was found around the shores of Lough Neagh and the streams and rivers which run into it. It is also recorded in Galway, Fermanagh, and the Garron Plateau of Co. Antrim.

In 1939 it was found on the island of Coll in the Hebrides and subsequently on the island of Colonsay. Since then a total of at least seven mainland sites have been recorded in north-west Scotland, in Argyll and Inverness. Even more astonishing was the discovery of a colony flowering in Devonshire, while more recently it has been found in two fresh island sites in the Inner and Outer Hebrides.

The roots are long and fleshy, tending to spread horizontally in wet ground, and the stem is 10 to 30 cm in height. The leaves are long, narrow, and erect, nearly flat in the southern Irish plants but markedly curled back in the more northern plants so that they appear tubular and even narrower than they are.

The stem leaves are narrow and sheathing, and like the stem itself rather yellowish. The leafy bracts are long and erect and sheath the ovary.

The spike bears about twenty flowers in three distinct spiral rows, which differentiate the species from both the other *Spiranthes* and give the spike a compact appearance. The flowers are relatively large, creamy white, and have a broader, more showy labellum. All the floral parts are covered with glandular hairs.

The flowers are fairly strongly scented, the smell being described as like vanilla or hawthorn. On Coll one of the islanders described the scent as being so strong that, in a good flowering year, the plants could be detected at some distance on a warm, still day, even when they were out of sight behind a wall.

Insect-pollination is presumed, but has never been proven. The species is very erratic in its flowering, being present in large numbers in some years and scarce in others.

The Irish Lady's-tresses is a plant of marshy meadows and wet open habitats, and is always to be found near streams or rivers, or on the margins of lakes. Associated plants are Purple Moor-grass, Devil's-bit Scabious (*Succisa pratensis*), Meadowsweet (*Filipendula ulmeria*), Marsh Cinquefoil (*Potentilla palustris*), Ivy-leaved Bellflower (*Wahlenbergia hederifolia*), and *Spaghnum* mosses of the green type. It is never found in a really acid bog vegetation.

In the mainland Scottish sites the Irish Lady's-tresses is found in two rather different types of habitat.

In the first it is found on the margins of rivers and freshwater lochs close to the water's edge, where the ground is plainly submerged for much of the winter. In one site it is accompanied by the Marsh Clubmoss (*Lycopodium inundatum*).

The second type of habitat is much drier, with rocky outcrops and heather, intersected by damp drainage ditches and gullies. Strangely, the majority of plants flower on the higher ground, and not in the ditches as one would expect. J. Raven has made the interesting observation that the greatest number of flowering spikes appear in the area where, during the winter, the cattle have been fed, the plants seeming to benefit while dormant from the treading and manuring. However, during the growth phase the plants are extremely brittle, and any such trampling would prove highly destructive.

The flowering period is from mid-July to the end of August, the Scottish plants being at their best in mid-August.

Some authorities have divided the British plants into two types. The southern Irish plants, designated *Spiranthes gemmipara*, are described as having flatter leaves and white flowers with a broad labellum. The northern form, *Spiranthes romanzoffiana* var. *stricta*, is said to have narrower folded leaves and creamy-coloured flowers with a narrower labellum. However, studies in North America by Professor Oakes Ames of Harvard have shown that the two forms are not distinct but that in the normal population both extremes and all the intermediate forms will be present.

Listed in the *Red Data Book*.

16 Common Twayblade

Listera ovata (L.) R. Brown

The Common Twayblade is most aptly named as it has an extensive distribution throughout the British Isles (with the exception of the Shetlands) and is abundant in many of the places where it is found.

While many people dismiss it as a dull and uninteresting plant, a careful examination will reveal that it is not without charm and has the added attraction that it keeps good company, being a close companion of nearly all our most beautiful and rare orchids.

The fibrous roots grow in a matted cluster, and from them arises the strong stem, densely covered in glandular hairs. The normal height range is 20 to 60 cm but plants up to 75 cm have been recorded. There are two to three basal sheaths to the stem.

The dark-green, oval leaves are large and flat, and are borne some distance up the stem. Normally there are two opposite leaves—hence the name Twayblade—but rarely there may be a third leaf above or below the pair.

The flowering stem bears about twenty flowers, occasionally as many as a hundred, the bracts being barely half the length of the flower stalks. The flowers are yellow green, the short perianth segments clustering into a partial hood, occasionally tinged with reddish brown. The labellum is long and green, and tends to fold back beneath the flower. It is divided almost to midway, the two lobes so formed being rounded and slightly divergent, while the base is marked by a central groove which secretes nectar. The ovary is globular and slightly ridged.

Pollination is effected by small flies, ichneumons, and beetles. The rostellum is very fragile and explodes on contact, gluing the pollinia on to the insect's head and frightening it off to another flower. After the pollinia have been removed the rostellum hinges up to expose the stigma, the time taken for this to occur ensuring cross-pollination. Bees certainly visit the flowers for the nectar they secrete, but are too large to act as pollinators.

A high proportion of the flowers produce ripe seed capsules. Four years elapse before the young plant develops its first leaf, and a further ten years before it flowers. The Common Twayblade can also multiply vegetatively by developing buds on the rhizomes, an efficient method which can be carried on in dense woodland too dark to allow flowering. At the bottom of densely overgrown chalk-pits I have found many a flourishing colony which never flowers.

The Common Twayblade is a most adaptable orchid and can be found in a vast range of different habitats. It flourishes in moist woods on base-rich soils where it may even be the dominant ground plant. It will also be found in large numbers in the hill pastures of the north of England, or in the dune slacks of North and South Wales and the Isle of Man. In the Orkneys it grows on acid mineral soils with grass and heather, and it is also widespread in Ireland.

The plants growing in well-shaded woodland tend to be taller and darker green than those in exposed areas such as downland and dunes.

The flowering period is long, ranging from the first week in April in the south of England to the end of July in northern areas such as Northumberland and Teesdale.

No hybrids are recorded, but abnormal plants are not infrequent. These may show the three petals all shaped like the forked labellum, or conversely the labellum may resemble the other two short simple petals.

17 Lesser Twayblade
Listera cordata (L.) R. Brown

The Lesser Twayblade has all the characteristics of the genus *Listera* (the paired leaves borne some way up the stem, globular ovary, and flowers with a markedly forked labellum), but everything about it is tiny, and the plants are most elusive and difficult to find.

The roots are fibrous and the rhizomes slender and creeping, penetrating through the damp moss cushions in which it grows. These rhizomes are heavily infected with mycorrhizal fungus at all stages of the plant's growth.

The stem is from 3 to 25 cm in height, usually less than 5 cm, often reddish coloured, and covered in fine hairs.

The two leaves are small and heart shaped, shining apple green on top and paler underneath. The edges of the leaves tend to be wavy, and there is a well-marked central rib. The paired leaves of the non-flowering plants bear a very confusing resemblance to small plants of Bilberry, the two often growing together.

The stem carries three to fifteen flowers on short stalks, the bracts being tiny and insignificant. The colour of the flowers is always reddish, but may vary from pale pinkish brown to a deep copper, and usually all the floral parts are the same colour.

The ovary is fat and globular, and all the perianth segments except the labellum are blunt and spreading. The labellum is long and slender, projecting slightly forwards and boldly forked to above the middle, each lobe being sharply pointed. In some individuals there may be a pair of tiny lateral lobes at the base of the labellum.

The distribution of *Listera cordata* is northern, coming south as far as Derbyshire, Shropshire, and Cheshire, and in Wales as far as Brecon. It is also known from a small area of Devon, and from south

Somerset. It is widely recorded throughout Ireland. It has never been recorded from south-east England in the past, but W. Ingwersen recently found it in flower in Sussex. Further investigation may show whether this is the result of accidental introduction, or whether it has been there all the time and has simply been overlooked. There is no mention of it in a flora of 1871 which covers the area where it was found.

The Lesser Twayblade grows in two types of habitat: on open moors and bogs among Bell Heather and Ling, and in pinewoods. Since it is to be found in wet moss cushions underneath the heather, the microclimate is virtually the same in either case, and the difference of habitat not so marked as at first appears. In most cases the soil is acid and the moss very wet, the sites often being north facing.

I have found it growing on the sandy beach of a lochan in Wester Ross under clumps of heather, and in Ribbledale, Yorkshire, in rather dry moss under stunted Juniper bushes on a limestone pavement. The finest colony was in Gwynedd, where we made the first definite county record for the Lesser Twayblade. A previous record had been on the boundary of three counties in an area difficult to pinpoint. Here it grew in profusion in *Sphagnum rubellum* under heather and Bilberry in an area of tumbled rocks high up in the Rhinogs, the whole region being saturated with mist and rain. Since then it has been found in two quite different areas of the county.

The flowering period is from mid-May until the end of June, rather earlier than the time quoted in most books. The confusion may arise because the flowers do not appear to wither even when seed has set and dispersed. Thus one may find flowers still 'out' in July, when a careful examination of the ovary will show that the carpel valves are open and the seed has already gone.

Pollination is carried out by small flies and ichneumons, as in the case of *Listera ovata*, but self-fertilization may also occur, seed being set most efficiently.

In any colony a large proportion of the plants will be non-flowering—usually around 80 per cent—but it is very easy to overlook the tiny pairs of leaves if there is no flowering spike.

18 Bird's-nest Orchid

Neottia nidus-avis (L.) Richard

The Bird's-nest Orchid is a most strange-looking orchid with reduced leaves and little chlorophyll, flowering in the darkest parts of woods of beech and yew, where it is often the only flower in what is otherwise a monotonous carpet of dead leaves.

The true rhizome is concealed by the tangled mass of short fleshy roots, which give the subterranean portions of the plant the appearance of a badly made bird's nest. Both roots and rhizomes carry a heavy mycorrhizal infection, for the plant is wholly saprophytic and relies on the products of the rotted beech-leaf humus.

The stout stem is 20 to 50 cm high and slightly downy. The stem and all the parts of the flowering plant are a warm honey-brown colour. There are no true leaves, but the stem is covered with many brown scales and its upper parts carry several large sheathing scales like primitive leaves. The bracts are pointed and papery, and are shorter than the ovaries.

The single flowering spike is robust and dense, carrying fifty to a hundred fairly large brown flowers.

The perianth segments are short and form a loose hood above the broad and long labellum. The labellum is forked, the two lobes of the distal portion being rounded and divergent, and in addition it may have two small lateral teeth at the base. The base of the labellum forms a shallow cup which secretes nectar attractive to insects, this cup being in essence a primitive spur.

The flowers have a pleasant scent of honey and attract many sorts of small flies and thrips, which probably act as the pollinators.

The pollen masses are bright yellow and extremely friable, the explosive action of the rostellum attaching them to the insect visitors in the same manner as for *Listera*. Self-fertilization also occurs, and has been reported to occur in a remarkable manner by N. Bernard. He observed that some flower spikes fail to emerge from the ground as the result of some obstruction. These flowers have been known to set seed without the flowering spike ever appearing above ground.

The flowers of the Bird's-nest Orchid may open as early as the end of April, but the main flowering period is in June. Seed production is highly efficient and is the main way by which the species can spread, plants taking about nine years to reach maturity from seed.

The Bird's-nest Orchid is most common in the south of England, the distribution further north and west being widespread but rather patchy. In the south it grows under beech and yew trees, sometimes accompanied by another saprophyte, the Yellow Bird's-nest (*Monotropa hypopithys*). In Ireland it is associated with ancient oak- and mixed oak- and birchwoods which may be relics of the days when the forest cover was more extensive.

It is never abundant even in favoured areas, normally growing in twos and threes, although I have found over a hundred spikes in one small area of beechwood. The dead spikes of previous years are very persistent, often standing for more than a year.

There are no hybrids, but rarely plants may be found with flowers which are whitish or sulphur coloured.

19 Creeping Lady's-tresses

Goodyera repens (L.) R. Brown

To be sure of seeing this charming orchid it is necessary to visit Scotland, where it flourishes in certain parts of the ancient pine forests of the central and east Highlands.

The plants of Creeping Lady's-tresses have no tubers, but fleshy roots and stoloniferous rhizomes, which ramify through the rotted pine needles and the moss in which the orchid grows.

The hairy stem is 10 to 20 cm high, the leaves stalked, oval, and marked by a conspicuous net of veins. Each stem bears about twenty flowers in a single spiral row, but the spiral pattern is masked by the habit of the flowers turning to face the same direction. The bracts are longer than the ovary, sheathing at their bases, and ending in a long sharp point which is usually curved over at the tip.

The flowers are creamy white, short, and fat, with blunt perianth segments densely covered in glandular hairs. The labellum is spout shaped and pointed, with a slightly saccate basal portion which secretes nectar. The flowers have a sweet scent which attracts pollinating humble-bees.

Cross-pollination is ensured by a mechanism similar to that in *Listera*, except that in *Goodyera repens* it is the labellum which moves downwards to expose the stigma after the pollinia have been removed.

Cross-fertilization is usual, and the setting of seed is efficient. Seed does not play much part in the multiplication within a colony, serving chiefly to establish new ones.

The Creeping Lady's-tresses grows in the ancient pinewoods which are probably the relics of the great Caledonian Forest which once covered vast areas of the Highlands. Here the ground is covered by a deep layer of rotting pine needles and moss, which is very moist and friable. Through this the stolons ramify fairly close to the surface. After about five years they develop leaves, and at about eight years the stem develops several thicker fleshy roots. Having thus prepared a store of food the stolon turns upwards to produce the flowering stem.

After flowering, the main stem dies and the various side branches continue to grow as separate individuals. This process of vegetative propagation is the means by which the members of the colony multiply.

The flowering period is during July and August, the orchids being at their best in the latter part of July.

The main distribution area is in the central and east Highlands, especially the Spey Valley. There are other scattered locations in Scotland, but it no longer grows in the Orkneys, and has never been recorded in the Western Isles, where suitable habitat is lacking. It has been recorded as far south as Durham and Westmorland.

There remains a curious centre of distribution in north Norfolk where it has long been recorded in plantations of pine trees in a number of localities. In some of these sites it is thriving in considerable numbers. It has been assumed that all these colonies are the result of plants introduced with the soil around seedling pine trees from Scotland, but recent work suggests that at least some of these plants may be truly indigenous stock.

No hybrids or variants have been recorded.

20 Bog Orchid

Hammarbya paludosa (L.) Kuntze

[*Malaxis paludosa* (L.) Swartz]

The Bog Orchid is tiny, all green, and remarkably elusive, a character compounded of its minute size and erratic flowering habits. Although scattered throughout the British Isles, it is a rare plant and only to be found in any numbers in Scotland and the New Forest.

The roots are virtually absent, being reduced to root hairs, which have a very heavy mycorrhizal infection to cope with the plant's nutritional requirements.

At the base of the stem are two pseudobulbs, placed one above the other, which are formed by swellings of the stem. The prominent lower one bears the dried-up remains of the previous season's leaves, while the upper pseudobulb is enveloped by the bases of the new leaves.

The aerial stem is 3 to 12 cm high with one or two leaves reduced to sheaths at the base and two to four small oval leaves borne near the middle of the stem. The margins of the upper stem leaves form tiny bulbils, which detach and form new plants, a means of vegetative propagation.

The stem is five angled and carries as many as fifteen tiny green flowers, the bracts at the base of each flowering stalk being very small.

The flowers of the Bog Orchid are curious in structure, since the labellum is uppermost. In most other orchid species the labellum lies ventrally, having rotated through 180° from its primitive position. The Bog Orchid is unique in that the rotation continues through 360° so that the labellum again points upwards.

The flowers are so minute that it is very easy to misidentify the floral parts, but Figure 7, which magnifies the flower ten times, should make it clear. The ventral structure is a large sepal, the other two sepals pointing upwards and almost enclosing the triangular pointed labellum. The labellum bears longitudinal stripes of alternating light and dark green. The other two petals are small and lie laterally—they may even fold round the back of the flower.

Cross-pollination occurs with regularity, the assumption being that it is performed by the small flies which are plentiful in the bogs. The flowers often set large amounts of seed, which is fine and floats on water, but its fate is unknown. The production of the leaf bulbils obviously plays an equally important role in the propagation of the species.

The Bog Orchid is only found where there is a good growth of *Sphagnum* moss, although it will grow in among reeds (*Phragmites*), provided that their growth is not so thick that the moss cushion is

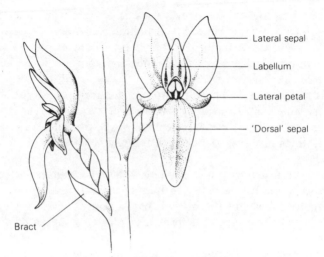

Lateral sepal

Labellum

Lateral petal

'Dorsal' sepal

Bract

7 Enlarged flowers of the Bog Orchid, *Hammarbya paludosa* × 10

reduced. The habitat is always saturated with water which is often acid, although not extremely so. Certainly this is true of the New Forest sites where the Bog Orchid still flourishes.

Elsewhere it is a vanishing species, as reference to many an old flora will show, since many of the extensive bogs in the south of England have been drained to extinction. Richard Deakin, in his *Flowering Plants of Tunbridge Wells and Neighbourhood* (1871), writes: 'Press-bridge Warren, near Wych Cross, Ashdown Forest; abundant on the great bog near the Kidbrook Park pales. In a bog near the windmill on Frant Forest.' The Bog Orchid has long since disappeared from all these sites.

The Bog Orchid has a wide western distribution in Ireland, Scotland, and Wales, where it may be found quite high up in the mountains. It still clings on in west Norfolk, and has been recorded since 1980 in Northumberland, Durham, Breconshire, and North Devon. The plants are very small and difficult to find.

The Bog Orchid is notorious for its erratic flowering habits—one year it may be abundant, while a careful search in subsequent years may fail to find it.

The flowering period extends from July to September, being earlier in the south of England and in sites at low altitude. There are no varieties or hybrids.

Listed in the *Red Data Book*.

21 Fen Orchid

Liparis loeselii (L.) Richard

Apart from one old record for 1802 near Sandwich in Kent, the Fen Orchid is restricted to two main areas, East Anglia and South Wales, with an outpost in Devon on the opposite shore of the Bristol Channel. The two populations have their own distinctive characters, developed possibly in response to the very different types of habitat in which they are found.

The Fen Orchid has a thick root-stock, to which are fixed a pair of very hard pseudobulbs developed from the base of the stem, the old pseudobulb being surrounded by the bases of the old leaves.

The stem is 6 to 20 cm in height, the upper part being markedly three sided, the angles extending at the tip into narrow wings. The two leaves are bright yellow green, pointed oval in shape, and overlap at their bases where they sheath the new pseudobulb. The leaves of the East Anglian plants are about four times as long as they are broad, and rather elegant in shape.

The stem carries four to eighteen yellow-green flowers. The perianth members are long and narrow, while the labellum is shaped like a broad spearhead, and is folded into a gutter. The labellum may point upwards like that of the Bog Orchid, but this is not a constant feature and in any group of plants flowers will be found where the labellum is variously directed.

In East Anglia the Fen Orchid is now restricted to the fens of Cambridgeshire, Norfolk, and Suffolk, where the water is either alkaline or neutral. The Fen Orchid in its natural environment appears at a certain stage of colonization of pool edges with reeds (*Juncus subnodulosus*), sedges (*Schoenus nigricans*), and moss. It is thus a plant of a changing situation, which may be kept flourishing by artificially controlling the environment. In these same areas will be found the Early Marsh-orchid, sometimes in the rare yellow form *Dactylorhiza incarnata* ssp. *ochroleuca* which is peculiar to the fens, also the Irish

75

Marsh-orchid, Pennywort (*Hydrocotyle vulgaris*), and various mints (*Mentha* spp.).

It has been found that successful flowering follows the cropping of the reeds, plants flourishing the following summer in the areas where the reeds have been cut. In the succeeding summers the number of flower spikes decreases as the reed growth returns, plants probably remaining in a non-flowering state for many years. Such plants would be very hard to find, as anyone who has struggled through the luxuriant growth of an East Anglian fen in summertime will readily acknowledge.

The Fen Orchid has become more scarce as fenland has been drained and cultivated, but now the fenland reserves where it grows are most carefully managed, so that optimal conditions will be created by a programme of reed cutting in selected areas.

The form in South Wales, var. *ovata*, is dumpy in habit with very broad leaves only one and a half times as long as they are wide and relatively shorter perianth segments. It is found in the slacks of stabilized dunes among Creeping Willow (*Salix repens*), the Early Marsh-orchid, Southern Marsh-orchid, Marsh Helleborine, and some *Sphagnum* moss, these areas being fairly moist for most of the year. Plants of this squat, broad-leaved form should in theory be more resistant to drought, but I found that in a colony of over 300 plants only three flower spikes had not shrivelled in the constant blast of salt-laden wind. Despite this, the colonies appear to be flourishing.

In early life the Fen Orchid is largely dependent on its mycorrhizal fungus, but later, when the normal roots develop, the plants become independent. The pseudobulbs are first formed in the second year, and the leaves first appear in the fourth year.

There is some argument as to whether cross-fertilization is the normal method of reproduction. M. J. Godfery observed that self-fertilization could not be the usual method since seed was not often set, while Continental field-workers maintain that the rostellum is so poorly developed that self-fertilization is easy and more or less inevitable. Seed setting appears to be efficient whatever the method; the ripe seed capsules are erect and in shape rather like the seed capsules of Long-headed Poppy (*Papaver dubium*).

Vegetative multiplication can also occur. In a good year two or more pseudobulbs may be produced, their separation leading to the establishment of new individuals.

The flowering period is from mid-June to the early part of July.
Listed in the *Red Data Book*.

22 Coralroot Orchid

Corallorhiza trifida Chatelin

The Coralroot Orchid is one of the three saprophytic orchids which
grow in Great Britain, but unlike the Ghost Orchid and the Bird's-nest
Orchid the plant does have some chlorophyll-containing cells in the
stem and the ovary. From these a small amount of food may be
manufactured by photosynthesis, although the plant is virtually
dependent on mycorrhizal fungus.

The Coralroot Orchid leads an entirely subterranean existence apart
from the production of flower spikes. The rhizome mass is composed
of creamy-coloured rounded knobs rather like coral, from which are
developed small tufts of hairs. There are no roots and unlike *Epipogium
aphyllum* there are no long runners.

The leafless stem is rather yellowish, 10 to 20 cm high, and has two
to four long sheathing scales. These scales are brown, often shrivelled
at the top edges, and extend half-way up the stem. In mature plants as
many as ten aerial stems may develop from one root-stock, so that
small clumps of flowering spikes are not infrequent.

Each spike carries four to twelve pendant flowers whose over-all
colour is yellowish or brownish green. The upper sepal and the two
upper petals form a hood and the two lower sepals curve downwards
and forwards around the base of the labellum. The tips of the sepals
and petals are brown, which makes the flower appear to be withering
when in fact it is at its best.

The labellum is white, much broader than the other perianth
segments, and folds down from its base. The edge is slightly frilled
and the base bears a number of bright crimson spots. A close
examination of the flower reveals a charm which is belied by the rather
dingy appearance of the whole plant.

It is not clear how important insects are in effecting pollination. The
rostellum is small and degenerate so that it is simple for the pollinia to
fall forward on to the stigma. However, the pollinia are easily detached,
the flowers secrete nectar, and I have seen them visited by hoverflies

and by dungflies of the genus *Scatophaga*. Seed is set in most of the flowers and, since there are no runners, seed must play an important part in spreading this species.

The distribution of the Coralroot Orchid is decidedly northerly; it is most abundant on the south side of the Moray Firth, in Fife and Angus, and in parts of Northumberland, Cumberland, and Westmorland. Since 1979 it has been found in five new Scottish vice-counties. It still exists in Yorkshire at its most southerly site in Great Britain.

It is a plant of mossy pinewoods of mixed pine and birch, where it may be accompanied by *Goodyera repens* and *Listera cordata*. It can also be found in damp areas of sand-dunes, even among Marram grass and in Alder carrs. I have found it on ledges in a quarry in stony ground under sallows, with Common Lady's-mantle (*Alchemilla vulgaris*), but wherever it is found the ground is always damp.

The flowering period is from early June to the end of July. The flowering spikes are very hard to see against a background of grasses and sedges, and, since nothing is visible above ground at any other time, this factor, combined with erratic flowering, may make the Coralroot Orchid appear to be rarer than it really is.

No variants or hybrids are recorded.

23 Musk Orchid

Herminium monorchis (L.) R. Brown

Many of our orchids have received their names on account of a fanciful resemblance to some animal, but the Musk Orchid rejoices in English and scientific names which are both founded in fallacy. The orchid is sweetly scented but does not smell of musk. Each plant bears a main tuber from which the flowering stem arises, and also as many as five smaller stalked tubers. If the plant is uprooted these fine stalks break, leaving their tubers in the ground and yielding an orchid which appears to have just the one tuber which its scientific name implies. Two or three tubers are produced each year on stolons up to 20 cm long, an efficient means of vegetative propagation which often leads to the formation of clumps of plants. There are also a few thread-like roots which may carry a mycorrhizal fungus.

The Musk Orchid is very small, 5 to 15 cm high. All the parts of the plant are a bright yellow green; the only orchid it could be confused with is the Bog Orchid which is never found in grassland. There are two or three short, strap-shaped basal leaves, and usually a single small pointed stem leaf. The bracts equal the ovaries in length.

The flowering spike is densely packed with twenty to thirty tiny flowers, their segments pointed and connivent so that they form minute spiky bells. In one colony I have often found plants 20 cm high with over a hundred flowers. The upper sepal is broader than the two laterals, each of which has two opposite teeth. The petals also bear a tooth on each side. The labellum is shorter, three lobed, with a long forward-pointing central lobe. At its base it bears a cup-shaped hollow which represents a primitive spur. The flowers smell of honey.

Each of the pollinia has a large round viscidium. The flower is so constructed that any visiting insect must enter sideways between the labellum and the lateral petals, thus contacting the viscidium which sticks on to a leg. Small flies and beetles act as pollinators, and the flowers can also be self-fertilizing. Seed is set abundantly.

The Musk Orchid is restricted to the south of England, being found on the chalk downs of Kent, Sussex, Surrey, Hampshire, Wiltshire, and the Chilterns. It occurs west to Somerset and Gloucestershire on the oolitic limestone, and at one locality in South Wales. In the past it was found on the chalk hills of East Anglia as far north as Hunstanton but has not been recorded there in recent years.

It requires very short turf to grow in and favours the edges of paths, earthworks, old quarries, and the spoil heaps of ancient chalk-pits. It seems to like full sun, and appears to be very resistant to drought. The numbers flowering in any one year can vary enormously, so that while in a prolific season the grass may be dotted with tens of thousands of spikes, in the following years only a few may be found. In one Sussex site there are always several hundred spikes each summer, but in 1965 and 1966 I counted up to 20,000 before I gave up from mental exhaustion. Being so diminutive they are easily overlooked.

It is a remarkably persistent species. It still grows in Kent in an old chalk-pit where Deakin recorded it in 1871, and at one Buckingham-shire site first noted in 1886 there were 234 spikes in 1964.

Early July is the best time to see the Musk Orchid in flower, the flowering period ranging from mid-June to early August.

No variants or hybrids are recorded.

24 Frog Orchid

Coeloglossum viride (L.) Hartman

[*Habenaria viridis* (L.) R. Brown]

The Frog Orchid has two features in common with the Musk Orchid—its green colour and small stature make it easily overlooked and it is difficult to understand how it acquired its name, since the flowers bear precious little resemblance to a frog.

The plant has two palmately divided tubers, a new tuber being developed each year, and from this the new aerial stem arises. Although the young plants are to some extent dependent on mycorrhizal fungus, the tubers when formed are free of infection.

The stem is 5 to 35 cm high, slightly angled, often reddish coloured in its upper parts, and bears one or two sheathing scales. There are three to five broadly strap-shaped basal leaves, and several more pointed stem leaves. The bracts are long, leafy, and pointed, the lower bracts often much longer than the flowers, and the upper bracts being equal to, or shorter than, the flowers. In some very large specimens the spike may carry more than fifty flowers.

The flowers are green, often tinged with brown, and in some cases the entire flower is rusty brown. The sepals and petals are short and broad, forming a neat hood, while the strap-shaped labellum hangs down or may be folded back under the ovary. The labellum has three lobes at the tip, the central lobe being shorter, and a short blunt spur at the base. The labellum is usually paler coloured than the rest of the flower, the brown coloration being more marked at the edges. The flowers have a faint scent of honey.

The yellow, club-shaped pollinia are easily detached by the small insects which act as pollinators. After removal they hinge forward in a similar manner to the pollinia of *Orchis mascula*, but the process is much slower, so that by the time they are suitably positioned to strike the stigma the insect has inevitably visited another plant. Odd pollinia can frequently be seen stuck to the sepals or labellum, having detached spontaneously or failed to stick to a visiting insect.

The production of seed is highly efficient, a necessary feature since there is no means of vegetative propagation. The plant leads a subterranean existence for two to three years before the first leaves are produced.

The Frog Orchid is widely distributed throughout Great Britain, especially in central-southern England, North Wales, and north England to the Borders, and again in north-west Scotland, the Hebrides, Orkneys, and Shetland, and in Ireland. It is nowhere common, being found in chalk and limestone pastures where the grass is short, on old earthworks, mountain pastures up to 1,000 metres, the damper areas of stabilized sand-dunes, and on the machair of the Hebrides. On the island of Tiree the Frog Orchid appears to have increased greatly from 1956, when only a single plant was recorded, to 1973, when there were over 1,000 plants in twelve different areas of the island. It is, however, rather erratic in its appearance, flowering in large numbers and then vanishing from an area for some years. It is also curious that it should be missing from areas where the habitat is apparently suitable. For example, it is very rare in Kent, yet relatively common on the Downs of Sussex.

The plants are fairly resistant to drought, but suffered severely in the summers of 1975 and 1976 in the south of England. I found several colonies where flower spikes were littered about as if they had been picked and then flung down. At first I thought it was just vandalism but then realized that hundreds of plants were involved over a large area. A close examination revealed that in each case the flowering stem had dried out just below ground level, the base being shrivelled and blackened. There was evidence that the spike had then flailed about in the wind until it eventually broke off and fell some short distance from the plant.

The Frog Orchid is in full flower in early June, the flowering period ranging from the extraordinarily early date of 26 May to mid-September. One plant which grew near Lewes in Sussex flowered every year from 1965 to 1972. It was a particularly large specimen, and always came into flower a fortnight earlier than any other plant in the colony.

Hybrids with other orchid species have been recorded not infrequently, and include the Fragrant Orchid, Lesser Butterfly-orchid, Common Spotted-orchid, Northern Marsh-orchid, and Broad-leaved Marsh-orchid (*Dactylorhiza majalis*).

The hybrid *C. viride* × *Dactylorhiza fuchsii* was recorded in Co. Down by J. Wilde in 1986. Photographs found recently, taken by

E. J. Bedford in 1917 of a plant sent to him by Col. Godfery, show what appears to be the hybrid *C. viride* × *D. praetermissa* from Hampshire.

25 Fragrant Orchid

Gymnadenia conopsea (L.) R. Brown

[*Habenaria conopsea* (L.) Bentham]

One of my earliest recollections of orchids is of a field full of Fragrant Orchids, each flower looking as if it were made of pink sugar-icing, while the air was richly scented with a smell like carnations.

The Fragrant Orchid is found throughout the British Isles in a variety of habitats and often in vast numbers.

The tubers are palmately lobed and divided to midway. The stem is from 15 to 75 cm in height, most of the grassland plants being 15 to 20 cm high, while the fen forms are much taller. At the base are three to five long narrow leaves which are folded and slightly hooded, while another two to three stem leaves are narrow and pointed and lie close to the stem. The bracts are green and pointed, being equal to, or slightly shorter than, the flowers.

The flowering spike is long and fairly dense, with as many as 200 bright pink, sweetly scented flowers. The lateral sepals are fairly long, parallel sided, and project downwards and outwards on either side of the labellum. The other perianth members are connivent in a loose hood. The labellum is short and broad, distinctly three lobed, and bears a very long, slender, down-curving spur. In many cases this is translucent so that the nectar within it can be clearly seen.

The pollinia are club shaped and are attached to separate viscidia. The flowers are attractive to moths, butterflies, and bees, and as many as seven pollinia have been found attached to the proboscis of a moth. The pollinia are initially attached in an upright position, but then swivel to point forwards so that they strike the stigma of the next plant visited. Bees play a less important role than do moths, since the proboscis is less well adapted for feeding from such a long-spurred flower, but they do visit Fragrant Orchids, and may account for some of the hybrids formed with shorter spurred species. There is also a most interesting link between bees and several species of crab spiders, which lurk in the flower spike and catch unwary bees. These spiders can be found in quite a number of Fragrant Orchids and also in some of the Marsh Orchids and, despite their rather vivid coloration, manage to conceal themselves with evident success.

A high percentage of the flowers set seed and an aerial stem is produced in the third year of growth. However, under certain circumstances the period taken to reach maturity may be much shorter, and Sipkes in Holland recorded *Gymnadenia conopsea* ssp. *frisica* flowering in three years from the distribution of seed.

The Fragrant Orchid is widely distributed throughout Great Britain. In the south it is particularly common on downland and limestone pastures where it often flowers in extensive colonies with Common Spotted-orchid, Common Twayblade, Bee Orchid, and Pyramidal Orchid. Although there is a degree of variation between individual plants, the flowers are pink with a fairly strong scent of carnations. The labellum is moderately broad, and the three lobes are almost equal in size. Flowering takes place in June and July, most plants being in full flower in the second week in June.

A second form has been recorded extensively in marshes and fens in eighteen vice-counties in southern England from Somerset and Sussex north to Anglesey. It has been given the name *Gymnadenia conopsea* ssp. *densiflora* on account of the massive, densely packed flower spike. The leaves of ssp. *densiflora* are broader than those of the normal form, and the flowering spike is robust, reaching a height of 75 cm. The flowers are rather more mauve than those of the southern form, with a strong scent of carnations, and have a broad labellum, the two lateral lobes being much larger than the central lobe. The flowering period is later by at least two weeks, and the flowers are at their best in July and August. This form corresponds to the *Gymnadenia densiflora* of Wahlenberg. On account of its fenland habitat, *Gymnadenia conopsea* ssp. *densiflora* is accompanied by various marsh-orchid species, especially *Dactylorhiza incarnata* and the Marsh Helleborine.

F. Rose in personal communication differentiates a third distinct form with a northern distribution. This form flowers later in July and August, in hill pastures which may even be acid, and at heights up to 700 metres, where it is often accompanied by the Lesser Butterfly-orchid, Northern Marsh-orchid, and more rarely the Small-white Orchid. It is a plant of moderate stature with small, dark-pink flowers,

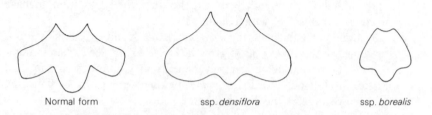

Normal form   ssp. *densiflora*   ssp. *borealis*

8 Forms of the labellum of the Fragrant Orchid, *Gymnadenia conopsea*

the lobes of the labellum being small and less distinctly divided. The scent is also distinctive, the flowers smelling strongly of cloves. This form, tentatively named *G. conopsea* ssp. *borealis*, is the type which predominates in the areas of Yorkshire, Durham, Teesdale, and Scotland where I have seen it. Rarely one encounters late-flowering Fragrant Orchids of this type in the south of England and I have seen them on the Downs of Sussex, while F. Rose has recorded them in the New Forest and on Ashdown Forest.

In view of the various differences in the characteristics of the three forms in their size, colour, scent, shape of labellum, and flowering date, it is worth considering whether they should be raised to specific rank of the degree accorded to the various marsh-orchids.

Albinism and partial albinism, which produces pale, almost white flowers, are common among the downland form of *Gymnadenia conopsea*.

A very curious sport was found on the Sussex Downs in 1970 by M. Ebbens of Gröningen in Holland. The plant was of normal stature and colour, but the three lobes of the labellum were long and pointed. The sepals were also long, broad, and pointed, thus giving a spurious symmetry to the flowers.

Hybrids with other species have been widely reported and include the Small-white Orchid, Common Spotted-orchid, Heath Spotted-orchid, Southern Marsh-orchid, Northern Marsh-orchid, Pyramidal Orchid, and Frog Orchid.

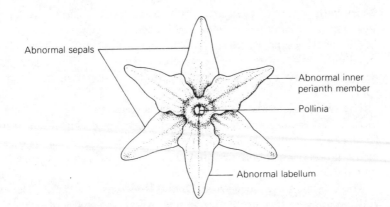

Abnormal sepals

Abnormal inner perianth member

Pollinia

Abnormal labellum

9 Abnormal form of the Fragrant Orchid, *Gymnadenia conopsea*, found by M. Ebbens

26 Small-white Orchid

Pseudorchis albida (L.) A. & D. Löve

[*Leucorchis albida* (L.) E. Meyer; *Gymnadenia albida* (L.) Richard; *Habenaria albida* (L.) R. Brown]

Smaller and less flamboyant than its close relative the Fragrant Orchid, the Small-white Orchid is a charming plant. There are some old records for it in the Ashdown Forest area of Sussex—it is described as being found in 1871 by J. J. Woods, Esq., near Nutley—but nowadays one needs to go to the north-west of Scotland to see it with much chance of success.

There are two small tapering tubers which are divided right to the base, the pointed tips functioning like roots. The true roots spread horizontally close to the surface of the soil, and carry a moderate mycorrhizal infection.

The stem is 10 to 40 cm in height, the majority of plants being less than 20 cm, with four to six fairly broad, flattened, oval leaves at the base and several small pointed stem leaves. The bracts are narrow and pointed, just longer than the ovary.

The flowers are small and bell shaped, borne in a fairly tight cylindrical spike. The spikes of most mountain plants bear twenty to thirty flowers, but I have seen some magnificent specimens in North Wales with over 70. They are white, with in some cases a tinge of yellow or green, and have a faint vanilla scent.

The sepals and petals are short, blunt, and connivent in a hood. The labellum is small, with three pointed divergent lobes, the centre lobe being longer. At the base of the labellum is a short, blunt, conical spur.

The vanilla scent and abundant nectar attract butterflies, day-flying moths, and solitary bees, whose visits, combined with some self-fertilization, result in ripe seed capsules forming in nearly 90 per cent of the flowers. Following germination an aerial stem is formed in the fourth year.

The habitat of the Small-white Orchid is hill pastures, grassy slopes, and mountain ledges up to 700 metres, where the soil is often rather poor and dry. It grows with equal facility in calcareous and non-calcareous soils, but they are always well draining and often in sunny positions. There it is found with the Fragrant Orchid, Lesser Butterfly-orchid, and the Heath Spotted-orchid.

Apart from odd records in the past for the Ashdown Forest ridge in Sussex, the Small-white Orchid occurs in a few scattered localities in north-west Wales, Derbyshire, Yorkshire, and Teesdale, and then with greatly increasing frequency throughout north and north-west Scotland. It is recorded from Skye and the Inner Hebrides, and in Ireland it occurs widely in most counties except those on the central plateau.

The flowering period stretches from the end of May to mid-July, depending on the area and altitude, but most plants are in flower in mid-June, the flowers tending to wither rather quickly.

Hybrids with the Fragrant Orchid have been recorded in Yorkshire and in several parts of Scotland, especially in Inverness.

∧
1
∨

2

3

5

ssp. *ochroleuca*

< **6** ∧

< **7** ∧

< **8** >

< **9** >

9a

10

< 11 >

< 12 >

< **13** >

∧ **15** <

< 16 >

< 17 >

< 18 >

< 19 >

< **20** >

∧
< **21**
>

var. *ovata*

< 22 >

∧ 23 >

<
24
>

∧
25
<

albino form

ssp. *borealis*

ssp. *densiflora*

< 25 >

< 25 >

< **30** ∧ ∨   var. *flavescens*

var. *trollii*

∧

30

<

∨

semi-peloric form

ssp. *jurana*

27 Greater Butterfly-orchid

Platanthera chlorantha (Custer) Reichenbach

[*Habenaria virescens* Druce; *Habenaria chlorantha* (Custer) Babington]

The two butterfly-orchids may have features in common, but it is an easy matter to tell between them once it is understood that there is no clear distinction on size alone.

The Greater Butterfly-orchid has two ovoid tubers shaped like small parsnips and a number of thin horizontal roots which penetrate the surface humus. The stem is 20 to 60 cm high, the base being surrounded by two to three brown sheaths, above which are the two large, elongated, oval leaves. The leaves are pale green, often with a slightly bluish tinge, and the upper surface is shiny. Despite their size the leaves are rather fragile and suffer from the depradations of slugs which chew large holes in them. Above the main leaves there are several small, pointed stem leaves. The bracts are long and pointed, equalling the ovary.

Each stem bears ten to twenty-five large white flowers, which stand well out from the stem on their long, slightly S-shaped ovaries. This gives the whole spike a broad appearance.

The two lateral sepals are long, bluntly pointed, and wavy edged. They spread outwards and slightly downwards. The upper sepal and the two upper petals form a very broad loose hood.

The labellum is long and strap shaped, pointing straight down, the tip often tinged with green, especially in plants growing in shady places. At the base of the labellum is a very long, curved spur, which projects right across to the opposite side of the flower spike, curving downwards to form part of a semicircle. The entrance to the spur is wide and clearly visible at the throat of the flower.

The pollinia are distinctive and serve as the most reliable way to differentiate this species from the Lesser Butterfly-orchid. They are borne prominently on either side of the wide entrance to the spur, and slope forwards and inwards so that they are closer together at the top. Each is attached to a round viscidium which is large and easy to see.

The flowers have a strong sweet scent which is reminiscent of freesias, lacking the rather sickly quality of the scent of the Fragrant Orchid. The flowers are more strongly scented at night and attract night-flying moths. I have also noticed that in dim light the flowers have a luminous quality. This would not appear to be luminescence, such as some fungi exhibit, and is probably just a characteristic of the colour of the flowers, but in the dim light of a woodland at dusk it must serve as an added attraction to visiting moths.

When the insect thrusts its head into the flower to suck the nectar from the long spur, the viscidia adhere to the side of the insect's head. After a short time the caudicle contracts and the pollinia swivel forwards and inwards so that they contact the centrally placed stigma of the next flower visited. Seed is set in 70 to 90 per cent of the flowers.

The Greater Butterfly-orchid is capable of remaining in a non-flowering vegetative state for many years until conditions change and encourage flowering. In one wood which I knew well hundreds of flowering spikes appeared in the year following the felling of a belt of mature trees. To my knowledge there had been no flowering Greater Butterfly-orchids there for twenty years, although they had been recorded there a long time previously and I had searched for them many times unsuccessfully.

The Greater Butterfly-orchid is widely distributed throughout south England and west Scotland, with less dense areas of distribution in central England, Wales, and Ireland. It is far more abundant in south England, where it favours base-rich or calcareous soils, especially woodlands lying on heavier soil below the chalk downs. It will flourish in scrubby damp woods where there is considerable shade with a mossy covering to the ground. There it is accompanied by White Helleborine, Common Twayblade, Fly Orchid, and Early-purple Orchid.

It also grows in permanent pastures with the Common Spotted-orchid and Fragrant Orchid and occurs with the Lesser Butterfly-orchid in some of the small hayfields of the hill country of Wales and north England. In one such meadow near Dolgellau, Gwynedd, there must have been 2,000 spikes of each species, well concealed in the lush grass. Where the moorland and mountains really start it is far less common, its place being taken by the Lesser Butterfly-orchid.

The Greater Butterfly-orchid flowers throughout June, a few early southern plants coming out at the end of May, while in the hills of Wales and Scotland they may still be in flower in the first week of July.

Woodland forms of the Greater Butterfly-orchid may be rather etiolated and delicate, so that care should be taken not to confuse them with Lesser Butterfly-orchids growing in the same locality. Examination of the flower structure and position of the pollinia will leave one in no doubt as to which species is involved.

Variants in flower structure are rare. A plant whose flowers each had three lips and spurs was found on the Isle of Skye in 1980.

The hybrid with *P. bifolia* has been recorded rarely. The supposed hybrid with *Pseudorchis albida* found in Perthshire proved to be a peloric *Platanthera chlorantha*.

28 Lesser Butterfly-orchid

Platanthera bifolia (L.) Richard

[*Habenaria bifolia* (L.) R. Brown]

The Lesser Butterfly-orchid occurs throughout Great Britain, but flourishes more in the cooler, wetter areas of the north and west, occurring in profusion in some hill pastures and on moorland.

The tubers are parsnip shaped like those of *P. chlorantha* and bear similar roots. The stem is 15 to 55 cm in height, the upper part being roughly triangular in section, and, like *P. chlorantha*, the base of the stem has brown sheaths.

The two base leaves of the woodland form are indistinguishable from those of *P. chlorantha*, being long oval in shape, while the stem bears a number of small bract-like leaves. The true bracts are narrow and pointed, not quite so long as the S-shaped ovary.

The stem normally carries fifteen to twenty white flowers in a loose spike, which appears narrower than the spike of *P. chlorantha* since the ovaries are not quite so long.

The general flower shape resembles *P. chlorantha*, the lateral sepals growing outwards and downwards on either side of the labellum, the upper sepal and two upper petals being connivent in a loose hood. The labellum is long and strap shaped and has a long spur at the base, which projects almost horizontally across the flower spike, without the tendency to curve down in a semicircle like the spur of the Greater Butterfly-orchid. Both the spur and the labellum are often tinged with green.

The most important distinguishing features are attributable to the narrow column. This means that the pollinia lie parallel and close together, so that the throat of the flower appears closed, while the viscidia are small and oval. These features are constant and entirely sufficient to differentiate the two butterfly-orchids.

The flowers are strongly scented, especially at night, and attract night-flying moths. When these attempt to feed from the nectar in the spur they brush against the viscidia which adhere to the base of the proboscis. After a short time the pollinia swivel to point forwards so that they strike the stigma of the next flower visited.

The distinct morphological forms of the Lesser Butterfly-orchid are found occupying very different habitats. The woodland form is most commonly found in the south of the country, especially in the beechwoods of south England. It is as tall as the Greater Butterfly-orchid but is not usually quite so robust and seems able to flourish in areas of deep shade. There it is accompanied by the White Helleborine, Common Twayblade, Bird's-nest Orchid, and Early-purple Orchid. It is a plant of elegant proportions and to see one lit by a shaft of sunlight in the dimness of a wood is unforgettable.

The moorland form is tubby in comparison, with a short stem and compact flower spike. The leaves are shorter, thicker, and more pointed, and tend to be folded so that they sheath the lower part of the stem. The flowers are rather creamy in colour and the perianth segments are shorter, so that the individual flowers appear smaller. This form predominates in the hill pastures and moors of the whole of the west of Great Britain, from Cornwall to the north-west of Scotland, as well as north and west Ireland. It is tolerant of quite acid conditions, growing among Bracken and heather, or in the damper areas near streams, where it is accompanied by the Fragrant Orchid, Heath Spotted-orchid, Early Marsh-orchid (*Dactylorhiza incarnata*), and Northern Marsh-orchid.

The woodland form comes into flower in late May and is at its best in the first three weeks of June, while the moorland form is still in full flower at the end of July in the hill regions of the north of Scotland and the Western Isles.

Variants are very unusual. C. B. Tahourdin records an unusual form in 1924 at Chilgrove in Hampshire, where the two upper inner perianth segments resembled the lower outer perianth segments. A spurless, peloric form was recorded at Calne in 1980 by J. H. Tucker.

The hybrid with *P. chlorantha* has been recorded, while a hybrid with the Frog Orchid was reported from South Uist, Outer Hebrides by J. Heslop-Harrison in 1949, but it has not been confirmed. Photographs taken by E. J. Bedford in the 1930s appear to show hybrids with *Dactylorhiza fuchsii* and *D. maculata* ssp. *ericetorum*.

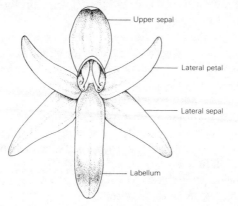

Upper sepal

Lateral petal

Lateral sepal

Labellum

10 Abnormal form of the Lesser Butterfly-orchid, *Platanthera biofolia*, as recorded by C. B. Tahourdin (from photograph).

29 Dense-flowered Orchid

Neotinea maculata (Desf.) Stearn

[*Habenaria intacta* (Link) Bentham; *Neotinea intacta* (Link) Reichenbach, f.]

The Dense-flowered Orchid was first recorded in 1864 in Galway by a Miss F. M. More, and for over a century thereafter it remained an entirely Irish possession, flowering only in the counties of Mayo, Galway, and especially Clare. There it is found in the limestone country of the Burren, where the extensive areas of limestone pavement harbour a unique flora.

Some of the plants recorded there are alpines such as the Spring Gentian and Mountain Avens (*Dryas octopetala*), for which the area is justly famous. Other plants such as the Strawberry Tree (*Arbutus unedo*), St Patrick's-cabbage (*Saxifraga spathularis*), Irish Saxifrage (*Saxifraga rosacea*), and the Dense-flowered Orchid all have their main centre of distribution in the Mediterranean.

91

One theory is that these plants are relics of a time prior to the last Great Ice Age or of one of the shorter interglacial periods during the main Ice Age, when the climate of western Ireland was much warmer than it is now. Somehow they survived when others elsewhere in Great Britain and northern France were exterminated by the ice and finally the community was isolated by the cutting of the links with the European land mass.

The tubers of the Dense-flowered Orchid are globular, crowned with a few fleshy roots. The aerial shoot develops in late autumn and persists throughout the winter, the flowering stem developing in the early spring. The stem is slender, 10 to 40 cm high, bearing two to three oval basal leaves and several elongated upper leaves which tend to sheath the stem. Some Mediterranean plants have leaves which are marked with rows of tiny purple-red dots, such plants having pink-tinged flowers, but most of the Irish plants have unspotted leaves and creamy-white flowers.

The flowering spike has a very odd appearance, since the flowers are small, scarcely open, and are borne on the end of fat, glistening, green ovaries more than twice their size. The ovaries are packed tightly around the stem in a rough spiral, but so arranged that the flowers all point in the same direction on one side of the spike. The bracts are narrow and less than half the length of the ovary.

The outer perianth segments form a tight, pointed hood, from which projects the short three-lobed labellum, the centre lobe of which is forked again and resembles the tongue of a snake. In some flowers this central lobe may be divided into three small teeth. The base of the labellum bears a short spur.

The flowers are creamy white, sometimes lightly tinged with pink, and are said to smell of vanilla. Most spikes bear fifteen to twenty flowers but I have seen large spikes with as many as thirty-two flowers.

The viscidia are contained in a pouch which is left behind when the pollinia are removed. Although the flowers are organized for cross-fertilization it appears that self-pollination is the general rule, the pollinia simply falling forwards on to the stigma. Seed production is efficient and plentiful.

Discoveries since 1980 have extended the Irish range of this species from West Donegal south to East Cork. It has long been associated with the limestone pavement and road verges of the Burren, but it has now been found in scrubby mixed ash and hazel woodland, the tall robust spikes more like those seen in Mediterranean colonies.

V. S. Summerhayes also records it as flowering in calcareous sea sand, and it was in just such a habitat that it was discovered in 1969 growing in the Isle of Man. The Manx colony is in stabilized sand-dunes with Marram (*Ammophila arenaria*) and Burnet Rose (*Rosa pimpinellifolia*), the plants proving quite extraordinarily difficult to see. They resemble most of the Irish plants in having unspotted leaves and nearly white flowers.

One can only surmise that they have come from the seed of the Irish orchids, but whether wind blown or delivered by sea can only be a matter for conjecture.

Both the Irish and the Manx plants are in full flower around 20 May, the flowering period being very short indeed. The high winds, which so often blow in both areas in May, lead to very rapid withering of the opened flowers, if they have not already been eaten by rabbits, who apparently find them delicious.

No hybrids have been recorded.

30 Bee Orchid

Ophrys apifera Hudson

The Bee Orchid is probably the best known of all the orchids of Great Britain. Its wide distribution and exotic colouring make this attractive plant much sought after, and in some measure this has contributed to the reduction of its numbers over the last twenty years. Sadly, many people still pick it under the misapprehension that the plant will flower again in another year but unfortunately the Bee Orchid is normally monocarpic and, having once flowered, the plant usually dies. However, some recent work done at St Christopher's School, Burnham-on-Sea, and recorded in *Watsonia*, has shown that plants may flower for eight consecutive seasons.

The flower spike ranges in height from 15 to 50 cm and arises from a basal rosette of five to six greyish-green, strap-shaped leaves. The tips of the leaves are often scorched and withered by midsummer. There are two sheathing stem leaves and the bracts are long, leaf-like, and exceed the ovary.

The spike bears two to seven flowers, the pink sepals each bearing three longitudinal green veins which are prominent in bud. The two upper petals are half the length of the sepals, brownish in colour, and appear cylindrical as their margins are rolled.

The red-brown labellum is three lobed, the central lobe being convex, velvety in texture, and marked with irregular bands of darker brown and gold at its base. The upper part of each lateral lobe forms a furry hump. Leaf-cutter bees have been observed to remove labella for nesting material.

The long column is beaked and resembles a miniature duck's head—the round, yellow pollinia lie under it, their long caudicles running in a pair of grooves.

The Bee Orchid takes five to eight years to reach maturity. If the weather is dry and windy at the end of May, mature plants may fail to flower, but produce a spike in a subsequent season.

Although adapted by mimicry for cross-pollination by bees, it is usually self-pollinated. A careful examination of the individual flowers will reveal that in nearly every case the pollen masses have swung down on their stalks, to fall straight on to the stigma.

This can be clearly seen in the close-up photographs. The frequency of self-pollination explains the persistence of variants such as var. *trollii*.

Seed setting is efficient and most plants will be found to bear three or more ripe ovaries. The seed is fine and dustlike—it is estimated that each seed pod contains 10,000 seeds.

The Bee Orchid is widely distributed in England and Wales, becoming rarer near the border and in Scotland. In Ireland it is widespread, recorded from nearly every county.

Although it favours short chalk turf, such as is found on undisturbed downland, abandoned chalk-pits, and railway cuttings, I have found it in abundance growing on heavy clay beside a road in Sussex in company with Green-winged Orchid and Adder's-tongue (*Ophioglossum vulgatum*).

It also flourishes on sandy soils, especially the calcareous shell sand of stabilized dunes, where it usually grows along the upper margins of the moister areas of the dune slacks. In one site in the Isle of Wight huge plants over 60 cm high and bearing up to ten flowers grow in profusion in a wet alkaline flush at the foot of sandstone cliffs. Most of these localities are open and sunny but I have found it in several places in Kent and Sussex growing in the depths of beech- and ashwoods on chalk and in Norfolk growing along the edges of rides in pine plantations on sandy soil.

It is always uncertain and fluctuating in its appearances. One year there were thousands growing on an old earthworks in Dorset, while in subsequent years only one or two could be found. The flowering period in the south of the country is from the first week of June until mid-July although it tends to flower rather later in the northern part of its range. In parts of North Wales there is evidence over the last few years that it is extending its range.

Hybridization is rare but has been recorded between the Bee Orchid and the Late Spider-orchid. The hybrid *O. apifera* × *insectifera* has been known from Bristol since 1968, and is the only natural population extant in Europe.

Several well-known forms of the Bee Orchid exist. In one variant the red-brown colour of the labellum is missing, so that it appears a light sage green, while the sepals are usually white. This form, var. *flavescens* (Rost), is widespread although never common. Variety *trollii* is a curious and distinct form where the labellum is long and pointed, barred with brown and yellow, while the appendage is not reflexed, but sticks out at the point of the labellum like a sting. Christened the Wasp Orchid, it is known from Gloucestershire, Dorset, and several other localities.

In the third variant—the semi-peloric form—the labellum is replaced by a structure resembling a pink sepal. First recorded in East Sussex it was described by C. B. Tahourdin in 1924, after which it disappeared for many years until rediscovered in 1967 not far from its original site. It flowered over a small area for the next five years, although in 1975 and 1976 there was no sign of it or of any other Bee Orchids, as the area was scorched in the drought. It is intriguing that this genetic aberration could have persisted so long unchanged and undetected in an area frequently visited by botanists.

O. apifera ssp. *jurana* was recorded by Richard Laurence in 1984 in Wiltshire. In this type the upper petals resemble pink sepals.

31 Late Spider-orchid

Ophrys holoserica (Burm. fil.) W. Greuter

[*Ophrys arachnites* (L.) Rechard; *Ophrys fuciflora* (Crantz) Moench & Reichenbach]

The Late Spider-orchid has always been a Kentish speciality, restricted in recent years to the chalk downs near Wye and Folkestone, although in former years it was a little more widespread within the county. There are a number of old unconfirmed records for Surrey, Gloucestershire, and Suffolk, the most recent being a record for 1891 at Bere Regis in Dorset. There has also been a recent unconfirmed record for the Sussex Downs, where it was found in 1974 by a botanist who knew the plant well in Kent.

The flowers are most attractive, probably the largest and most richly coloured of all the *Ophrys* species in Great Britain, and they have suffered greatly in the past from being picked and dug up. For this reason the Kent Trust for Nature Conservation has taken active measures to preserve and warden all the known sites where it flowers.

The tubers of the Late Spider-orchid are globular, and from them arises the stem which is 15 to 55 cm in height. The three to five basal leaves are arranged in a rosette. They are strap shaped with pointed ends and well-marked parallel veins, the upper side of the leaf being shiny. The two to three stem leaves are narrow and pointed, sheathing the stem. The bracts are like the stem leaves in shape and just exceed the ovary in length.

Most flowering spikes carry two to six flowers, although I have seen plants at Wye with as many as eight, and on the Continent plants with fourteen flowers have been recorded. The flowers are large and well spaced.

The three sepals are broad with rounded ends, bright pink with a prominent dark-green central vein. This vein shows up clearly when the flowers are in bud. The two upper petals are much shorter, somewhat triangular in shape, orange pink, and distinctly hairy, quite unlike the narrow, rolled upper petals of the Bee Orchid. Between them projects the rostellum which is quite prominent and green tipped, but nothing like as long as the duck's head-shaped structure in the Bee Orchid.

The labellum is very large and more or less square. The basic colour is chestnut brown, with a narrow paler margin, the whole labellum being velvety in texture except for the two furry humps on either side at the base. The upper part of the labellum is marked with a complex radiating pattern of creamy lines, which form a collar around the base of the labellum and tend to outline a group of dark-brown circles. This pattern is very variable.

The bottom margin of the labellum is broader than the base, and bears a bright yellow appendage in the centre. This appendage sticks out either downwards or forwards and is never neatly reflexed behind the labellum as in the Bee Orchid.

On the Continent the Late Spider-orchid is pollinated by male bees of the species *Eucera tuberculata*. The bees attempt to copulate with the flowers, and in so doing remove the pollinia on their abdomina. This curious behaviour ceases when there are plenty of female bees on the wing. Pseudocopulation has not been proven in England, but hybrids with the Bee Orchid have been recorded and could only have resulted from some similar insect activity.

The Late Spider-orchid is a very rare plant even in East Kent and is restricted to very short chalk turf. It is to be found on the narrow ledges which form on the sides of steep downland slopes and the mounds of earthworks, and seems unable to flourish in areas of longer grass where the Fragrant Orchid and Man Orchid grow.

The Late Spider-orchid comes into flower right at the end of June, and continues in flower until mid-July.

Considerable variation has been recorded in the colour of the sepals, which vary from dark pink to nearly white. Similarly there is much variation in the shape of the labellum, in its markings, and in the size of the lateral humps. Plants have been recorded with double labella.

The hybrid with the Bee Orchid has been seen on a number of occasions. C. B. Tahourdin recorded it near Folkestone in 1924 and there are some excellent illustrations of the hybrid in V. S. Summerhayes's *Wild Orchids of Britain*.

Listed in the *Red Data Book*.

32 Early Spider-orchid

Ophrys sphegodes Miller

[*Ophrys aranifera* Hudson]

The Early Spider-orchid is a smaller, more compact plant than the previous species, flowering much earlier in the year as its name would suggest, and bearing a far greater resemblance to the fat, hairy spiders one sometimes finds in gardens.

There are a pair of globular tubers with a few fleshy, shallow-growing roots. The height of the stem is usually 10 to 20 cm, plants of up to 45 cm being recorded on the Continent, while in seaside localities in England they may scarcely reach 5 cm. Being basically yellow green in colour, such tiny plants are very easily overlooked, and in one Sussex locality they grow in profusion along a well-trodden path, miraculously surviving the passing feet.

The three to four leaves are short and broad, forming a loose rosette. They are grey green, with well-marked parallel veins, and the tips are often browned and scorched by frost. The two to three upper leaves, which clasp the stem, also bear prominent veins. The bracts are leafy, upright, and slightly longer than the ovaries.

The sepals are large, roughly oblong, and yellow green. Their margins are slightly wavy, and the tip of the upper sepal curves forwards over the top of the flower. The two upper petals are short, strap shaped, and a brighter yellow green, their edges often tinged with brown.

The labellum is round, convex, and a warm rich brown colour. The centre of the labellum is velvety in texture and marked with a prominent irregular H, which is blue grey and smooth. On either side of the base of the labellum there are furry brown humps, which may be small and virtually absent, or large and prominent, projecting outwards and continuing down on either side of the labellum almost to the tip. There is no appendage at the tip of the labellum.

The column is shaped like a bird's head, but stouter than that of the Bee Orchid. On either side of the column are two prominent pouches (thecae) which house the pollinia. These glisten with nectar, and look just like a pair of eyes. They are obviously attractive to insects, since hybrids have occurred with other *Ophrys* species, but the percentage of ripe seed capsules set is poor, ranging from 6 to 18 per cent.

The Early Spider-orchid is nowadays only to be found with any regularity in Kent, Sussex, and Dorset. It has been recorded in Hampshire, Gloucestershire, the Isle of Wight, Berkshire, Oxfordshire, Northamptonshire, Cambridgeshire, and Suffolk. There is even an extraordinary record for Denbighshire at the end of the last century. In May 1988 F. Rose rediscovered it in south Wiltshire—a single plant.

It is a plant of short turf of old chalk downs and calcareous pastures, often growing within sight and sound of the sea. Recent work by M. Hutchings has shown that plants have a shorter life-cycle than previously thought. Few flower for more than three years, most emerging for one year only, before becoming dormant or dying. Almost all recruitment is by seed. Winter grazing produces a suitable short turf, but it is essential to protect the plants during the period of flowering and seed dispersal.

The Early Spider-orchid comes into flower in late April and early May, some plants remaining in flower until the end of the latter month. The flowers wither and fade to a pale buff colour rather quickly, so that one never finds all the flowers on a spike fully out at the same time, and they are rather susceptible to damage by frost. The numbers flowering in a locality may vary greatly from year to year.

There is considerable variation in lip shape, flowers showing circular and long pointed labella coexisting on the same spike. Some of the Dorset plants are of minute size, the labellum marked with a broad white inverted horseshoe which is veined with green. At one Kentish site some of the flowers had dark-brown sepals.

Physical variations are not infrequent. In Dorset I have seen plants with double lips and columns and inverted flowers, while in one Sussex colony several plants had flowers with conjoined labella and lateral sepals, the upper petals being much reduced, the whole flower smooth and suffused with green.

The hybrid *Ophrys insectifera* × *sphegodes* has been recorded several times in East Kent.

Listed in the *Red Data Book*.

33 Fly Orchid

Ophrys insectifera L.

[*Orchis muscifera* Hudson]

Although the Fly Orchid is the tallest of the genus *Ophrys* it is the most easily overlooked. Since it is frequently to be found in woods the leaves are often hidden by other plants, while the flower spike is long and spindly, the well-spaced small flowers being very hard to see.

There are a number of shallow-growing, fleshy roots above the egg-shaped tubers, from which arises the slender stem 15 to 60 cm in height. There are usually three lower leaves which are long and fairly narrow, broader in the distal third. They are dark green, floppy, and very shiny on the upper surface. The upper stem leaf is narrower, erect, and partly clasps the stem. The bracts are leafy, long, and erect, exceeding the ovary, which is also long, slender, and erect.

The flower spike is long with well-shaped, rather small flowers, most plants bearing two to ten flowers, although I have seen exceptionally large woodland plants with more than twenty flowers.

The sepals are pointed, yellow green, and stand out stiffly like the upper three arms of a cross. The two upper petals are wirelike, purple brown, and slightly velvety, having a remarkable resemblance to a pair of antennae. Between them is the short blunt column, which carries a prominent pair of pollinia. These lie closely side by side, the enveloping bursicle being orange coloured, their pale round viscidia clearly visible at the base of the column.

The labellum is long and three lobed, and quite the most beautifully coloured of any of the *Ophrys* species. The basic colour is a rich mahogany brown, the texture velvety, while across the middle is a band of brilliant, iridescent blue. This mimics to perfection the blue sheen which flies' wings show when they catch the light. There are two rounded, spreading lobes half-way down the labellum at the level of the blue band, and the tip of the labellum is divided into two shallow lobes. At the base of the labellum are two dark glistening patches which secrete nectar. Placed as they are immediately below the antenna-like upper petals they look just like eyes.

The whole flower is most convincingly insect like, and in fact is visited by male wasps of the species *Gorytes mystaceus*. M. J. Godfery noted in 1929 that the wasps exhibited pseudocopulation, this behaviour ceasing when there were plenty of female wasps on the wing. Despite this, the fertilization rate is poor, less than 20 per cent of capsules bearing seed. After germination growth is rapid, the first leaf appearing in the following winter and the first tuber during the next year.

The Fly Orchid is to be found mostly in south-east and south-central England. It becomes increasingly rare as one goes north, none being found recently north of Durham and Westmorland. There is an old record prior to 1930 for East Perthshire. In Wales it is found only in the extreme south, and then again in Denbighshire and Anglesey. In Ireland it is mainly a fen and lake margin plant, occurring from the Burren in the west to Kildare and Kilkenny in the east.

In south England it is found in beechwoods, especially the 'hangers' which are so characteristic of the slopes of the North and South Downs, on the edges of woodland, and in scrubby thickets. It tolerates quite heavy shade, flowering in bare earth with some mossy growth, or even through a dense ground cover of Dog's Mercury. Here it is accompanied by the White Helleborine, Narrow-leaved Helleborine, Greater Butterfly-orchid, Early-purple Orchid, Lady Orchid, Monkey Orchid, and Common Spotted-orchid.

The hybrid *O. insectifera* × *sphegodes* has been recorded several times from east Kent, and the rare hybrid with *O. apifera* from Bristol.

A number of variants in colour have been recorded, some flowers lacking both the brown and blue coloration of the labellum, which in consequence is greenish white. Flowers have also been recorded with double labella, or with all the inner perianth segments resembling the labellum—a flower with a three-bodied fly.

34 Lizard Orchid

Himantoglossum hircinum (L.) Sprengel

[*Orchis hircina* (L.) Crantz]

The earliest written record of the Lizard Orchid is for 1641 between Dartford and Crayford in Kent. There it continued to flower until about 1805 when it died out, for what reason we do not know. Until

1900 it remained a plant of extreme rarity, restricted to south-east England, but then it showed a remarkable change, being recorded in a large number of sites as far away as Yorkshire in the north and Devon in the west, most of these records being of solitary plants. It is possible that this spread, which climaxed in the 1930s, came in response to a change to a more oceanic climate, resembling that of north-west France where the species flourishes.

The Lizard Orchid is in all respects a large plant. The two tubers are broad ovoid in shape, and from them arises the stout stem 25 to 70 cm high.

The leaves are very large, oblong, and pale grey green, sometimes streaked and dotted with purple. They are formed in autumn, so that by the time the flowering spike arises the following summer, they are often withered and brown. Half the eight to ten leaves form a bulky, loose rosette at the base of the stem, while the rest loosely clasp the stem up to the base of the flowering spike. The bracts are linear and short, scarcely a quarter of the length of the twisted ovary.

The flowering spike is massive, and many contain more than eighty flowers, the whole structure looking most untidy on account of the long twisted labella which hang all round the spike.

The basic colour of the flowers is grey green, the sepals and petals being short, rounded, and formed into a close helmet. The inside of this helmet is marked with a series of parallel lines of brown dots and dashes, which just show through to the outside surface.

The three-lobed labellum is an extraordinary structure. The lateral lobes are long, brown, and crinkly, while the central lobe is 5 to 6 cm long, brown with a white base which is marked with bright crimson dots. The labellum is so long that when the flower is in bud it is coiled up tightly like a watch-spring. As the flower opens, this unfurls, at the same time twisting anticlockwise. The labellum thus looks just like the tail and two rear legs of a little lizard which has just taken a header into the flower—hence the name Lizard Orchid. The labellum bears at its base a short, round spur. The two pollinia are attached to one viscidium.

Apart from its bizarre appearance, the Lizard Orchid is remarkable for its smell, the flowers smelling very strongly of goats. I well remember finding a flowering spike which had been chopped off and was lying on the ground. Having decided to take it home for closer study, I very soon opened all the car windows. The flower finished the journey in the boot of the car and was promptly consigned to the garage, since none of us could bear it in the house. Quite how this aroma benefits the Lizard Orchid is not clear, but flowers are certainly visited by flies and bees. There is no proof of a regular insect vector for pollination but seed is set in about 30 per cent of capsules, and seed must obviously play a major role in disseminating the species across England.

The Lizard Orchid excites much attention by its unpredictable and sporadic appearance in areas where it has never been seen before. It may then flower for several years before disappearing again. This was the pattern of occurrence in the early part of this century, plants being found from Kent to Devon along the south coast, in Gloucestershire, Wiltshire, Oxfordshire, Bedfordshire, Cambridgeshire, Suffolk, Norfolk, and north to Lincolnshire and Yorkshire. One fine Bedfordshire colony flourished from 1938 to 1947 before being destroyed by ploughing. In Kent, Sussex, and Cambridgeshire it still flowers regularly and at one site in Kent as many as 200 spikes have flowered in one season. Elsewhere it has flowered recently in Devon, Wiltshire, Hampshire, Oxfordshire, Suffolk, and east Norfolk.

The Lizard Orchid is usually found in calcareous soils, often in areas of scrub and tall grass where it is accompanied by Fragrant Orchid, Common Spotted-orchid, and Pyramidal Orchid. Several sites are in old stabilized sand-dunes, similar to places where it is found in France and Holland. Here it is accompanied particularly by Pyramidal Orchid and Field Garlic (*Allium oleraceum*). It is quite extraordinary how difficult it is to see the Lizard Orchid against the background of a mass of the Garlic plants, into which they blend despite their size.

The Lizard Orchid flowers at the end of June and during early July.

No variants or hybrids have been recorded.

Listed in the *Red Data Book*.

35 Lady Orchid

Orchis purpurea Hudson

[*Orchis fusca* Jacques]

It is no crime to admit to having a favourite plant, and, as one born and bred in Kent, I have many happy memories of searching for this most regal of orchids in Kentish woodlands. The Lady Orchid was first recorded in 1666 in woodland near Gravesend, where it exists still. It is virtually restricted to chalk scrub and beechwoods on chalk in Kent, and may still be found in considerable numbers where it does occur.

There are two ovoid tubers 3 to 6 cm long, the current tuber fat and ivory coloured, that of the previous year being somewhat withered. Above these there are a few fleshy roots. The stem is stout, 20 to 100 cm high.

The leaves are large, broad oval, pointed at the tip, shiny, and dark green. The lower of the three to five leaves are arranged in a loose rosette, their bases sheathing the stem, while the upper leaves are narrower and clasp the stem. The bracts are very small, reduced to violet scales at the base of the glossy-green, three-ribbed ovary.

The flower spike is robust and may carry as many as fifty large flowers, which are shaped remarkably like little ladies wearing crinolines.

The sepals are formed into a compact hood which is dark reddish brown, the flecks of colour tending to form a pattern of horizontal lines. The two upper petals are small and strap shaped, completely hidden under the hood of the sepals.

The labellum is broad and flat, pale pink or white, and marked with a large number of tiny mauve or crimson spots. Close examination of these will show that they are tiny papillae, crowned with bunches of coloured hairs. The labellum has two narrow upper lobes which correspond to the arms of the 'lady', the lower part of the labellum being divided into two broader squarish lobes which often have frilled edges. Between the lower lobes there is a small triangular median tooth. The spur is pale pink and less than half the length of the ovary.

The Lady Orchid shows considerable variation both in colour and in the shape of the labellum. At one extreme one may find albino flowers with a white unspotted labellum and straw-coloured hood, while at the other end of the range of colour flowers will have a deep purple-brown hood and pink labellum heavily marked with purple. The shape of the labellum may similarly exhibit a range of variation. It may be dagger shaped, broad and undivided so that it resembles a spoon, divided into two arms with a single ventral lobe, divided into the customary two arms and two broad legs, and finally with all the lobes long and narrow and the central tooth prominent.

A whole range of colours and shapes may exist within one colony, but over all the Lady Orchid exhibits two fairly distinct forms which have separate areas of distribution. Those who wish to study the species in depth should read Dr Francis Rose's excellent paper on *Orchis purpurea* (F. Rose, 1949). In it he describes a west Kent form which is found on the downs on either side of the Medway Valley from Knockholt in the west to the Stour Valley in the east. The height is 20 to 38 cm with a dense flowering spike, ovary 1.3 to 1.9 cm long, and labellum blunt with well-marked spots. The East Kent form is found on the east side of the Stour Valley eastwards to Dover. The height is 30 to 76 cm, the spike lax, ovary 1.9 to 2.5 cm long, with a markedly lobed, less spotted labellum.

The Lady Orchid is very faintly scented and is visited by small flies which act as pollinators. M. J. Godfery in 1933 recorded flies of the species *Odyneras parietum* visiting flowers in Kent, and noted that in France bees also acted as pollinators. The percentage of capsules bearing seed is very low, from 3 to 10 per cent. The plants take eight to ten years to reach maturity, but then flower several times, although not necessarily in consecutive years. In any colony a high percentage of blind and immature plants will be found.

The Lady Orchid is a very persistent species, and I know of several sites where it has existed for well over a hundred years. Unfortunately, it is prone to damage by rabbits, which eat off the young plants with relish. This has certainly led to its disappearance from some of its old sites.

The Lady Orchid exists in over ninety sites in Kent, mostly in the east and north of the county, and has flowered in a couple of places in Surrey. There are old records from west Sussex, north Essex, and Herefordshire. In 1961 and 1964 B. Kemp found it in flower in the Thames Valley area of Oxfordshire. In 1986 R. J. Kemp refound a total of seven plants with three flowering spikes.

The Lady Orchid is restricted to chalk scrub or coppiced woodland on chalk, and beechwoods on chalk. In the latter it is often associated with yew trees and seems to like the bare terraces which form around the tree roots. The massive flower spikes set above the shiny leaves look superb against the background of dark yew foliage. There it is associated with Greater Butterfly-orchid, Fly Orchid, Early-purple Orchid, Stinking Hellebore (*Helleborus foetidus*), Spurge Laurel (*Daphne laureola*), and Gladdon (*Iris foetidissima*).

In the coppiced woodlands and scrub it is accompanied also by Common Twayblade, Primrose (*Primula vulgaris*), and especially Dog's Mercury. Such woodlands enjoy filtered sunshine and high humidity, and the Lady Orchid appears to flourish in these conditions. Rarely it is found in open grassland, usually near woodland colonies, such plants tending to be short and stocky.

The Lady Orchid flowers from mid-May to early June, but may be found in flower as early as the end of April in exceptionally warm years.

Apart from the variations in colour and shape which have already been described, F. Rose records a peloric form, where the two smaller upper petals assumed a structure like the labellum, resulting in a three-lipped flower.

No hybrids have been recorded in this country.

36 Military Orchid

Orchis militaris L.

In years gone by the Military Orchid occurred fairly frequently on the chalk hills of the Chilterns from Streatley in Berkshire through Oxfordshire and Buckinghamshire to Tring in Hertfordshire and Harefield in Middlesex. There are some old records for Kent, for Box Hill in Surrey in 1834, and more recently for a site near Goodwood in Sussex in 1924.

After 1886 it became progressively rarer, until by the late 1920s it was considered extinct in Great Britain.

Then in 1947 J. E. Lousley discovered a thriving colony in Buckinghamshire of thirty-nine plants with eighteen flowering spikes, in an area from which it had not previously been recorded.

An even greater surprise was the discovery in 1954 of a large colony in Suffolk, far from any previous record. The site was found during a routine survey prior to the compilation of the *Atlas of the British Flora* by the Botanical Society of the British Isles. Such routine checks of flowering plants are a time-consuming and painstaking task, but here was virtue amply rewarded!

The Military Orchid has two rounded tubers, the stem being 20 to 60 cm high and fairly stout. There are four to five large broad basal leaves, shiny on the upper surface, with well-marked parallel veins. Above these are one to two stem leaves which are smaller and sheath the stem. The pinkish bracts are small, less than a quarter of the length of the ovary.

Most flowering stems bear about thirty well-spaced flowers, but I have seen one superb spike in Buckinghamshire with fifty-seven flowers. The sepals and two upper petals form a large pointed hood which is much paler on the outside than within, the lilac or rose colour having a whitish cast as if coloured by pastels. The paleness of the flowers is very marked when they are in bud. The inner surfaces of the petals and sepals forming the hood bear a series of parallel lines which are mauve or dark green, the upper petals being completely masked by the sepals.

The labellum is long and divided into four lobes, which have a fanciful resemblance to the figure of a soldier, the spots on the labellum corresponding to rows of buttons on his tunic. The two lateral lobes which form the 'arms' of the soldier are long and narrow, rounded at the ends, curving outwards and forwards. They are usually tinged with lilac or pinkish mauve.

The body of the labellum is paler and quite long, the lower part being divided into two broad lobes which are squarish in outline, while between them is a small median tooth. The labellum is marked by a series of dark mauvish-red dots, which are formed by papillae set with tufts of coloured hairs, as in the Lady Orchid. These dots may appear to be formed in two lines, but in fact the spotting of the labellum is very variable, some plants having heavily spotted labella, while others are scarcely spotted. The labellum is pinkish, a colour which contrasts with the whitish mauve of the hood. The base of the labellum bears a rounded and slightly down-curved pink spur, half the length of the ovary.

The flowers have a faint vanilla scent, and insects are attracted to nectar which is secreted in the wall of the spur. This they obtain by piercing the spur wall with their mouth parts. In Europe the Military Orchid is visited by bees and humble-bees, while J. E. Lousley noted syrphid flies visiting plants in Buckinghamshire. I have noticed a number of flies of the species *Thricops semicinerea* visiting flowers in the Suffolk colony. Despite this, seed is set in very few of the plants.

The stem is first formed in the fourth year after seed is set, the plant first flowering in seven to eight years. Plants may then flower several times with intervals of two or more years between flowering, although I recollect one of the Buckinghamshire plants with odd narrowly lobed flowers which flowered in two consecutive years.

The Military Orchid flowers from mid-May to mid-June, on the edges of woods, clearings, or even rough fields where there is a fair amount of shade. The Buckinghamshire site is on a slope with yew trees, scrubby bushes, brambles, and Dog's Mercury. The plants have suffered in the past from timber operations, and more recently from human activity, the young plants being particularly susceptible to crushing underfoot. For this reason access has had to be strictly limited. It also occurs in the adjacent county of Oxfordshire.

The Suffolk site is in a shaded pit, the surface littered with stone fragments, but moist, with a moderate ground cover of moss under the scrubby privet bushes.

The Military Orchid flowers from mid-May to mid-June, the flowers usually being at their finest in the first few days of June. The flowering spikes may rival those of the Lady Orchid in size and are most impressive.

Within both colonies there is considerable variation in the amount of spotting of the labellum, ranging from heavily blotched, dark-coloured flowers to flowers with almost unmarked labella. One plant in Suffolk had all the lobes of the labellum long and narrow, with a prominent long median tooth, the flowers being deep lilac and virtually unspotted. Several other plants in the same colony had variegated leaves marked with alternate yellow and green longitudinal stripes.

In the past when the Military Orchid and the Monkey Orchid both flowered not uncommonly in the Chilterns, hybrids were noted on a number of occasions, as on a hill at Whitchurch near Pangbourne. This hybrid, and that between the Military Orchid and the Lady

Orchid, are most unlikely to occur nowadays, since the species are well separated geographically, although they occur in northern France where the three species may grow together.

Listed in the *Red Data Book.**

37 Monkey Orchid

Orchis simia Lamarck

First recorded in 1660, the Monkey Orchid was once a fairly common plant on the chalk hills of the Chilterns, mainly in the Thames Valley between Marlow and Wallingford. In that area it grew with the Military Orchid, with which it occasionally hybridized. This situation continued until the middle of the nineteenth century, when numbers declined catastrophically, probably as the result of increased ploughing of the old pastures, rapacious collecting, and the activity of rabbits. Between 1920 and 1923 five plants flowered in Kent between Canterbury and Dover.

There remained one good site in the Chilterns, with over a hundred spikes recorded between 1920 and 1930. In 1933, however, more than thirty spikes were picked. In 1949 the area was ploughed, although some bushes with the plants near them remained, but in the following year the whole area was again ploughed and the site destroyed. A few tubers were rescued and planted near by and, whether as a result of this or local seed dispersal, the Monkey Orchid persists in a small area of rough downland nearby.

In the 1950s there was a small but flourishing colony on the lawn of the vicarage in a village in north-west Kent. The vicar was in the habit of taking the ripe seed capsules and broadcasting the seed on the hills near by, where the species may yet reappear. This was last done in 1955. In 1956 there was one superb spike 38 cm tall bearing thirty flowers and leaves of a further six young plants.

On the retirement of the old vicar, the new incumbent would not undertake to safeguard the colony, so all the plants were moved with great care to private ground near by, which promised sanctuary and conditions resembling closely those under which it already grew. The largest plant flowered again in 1957 bearing forty flowers, which were visited by numerous small flies and one Large White Butterfly (*Pieris brassicae).* Since then the Monkey Orchid has disappeared from the site and no further records are known from West Kent.

Then in 1955 a single Monkey Orchid flowered in East Kent, only to be eaten by a horse. However, more plants appeared in the following year, and H. Wilks of the Kent Trust for Nature Conservation hand-pollinated them to ensure that seed was set. In 1957 plants flowered in the field site and in an adjacent wood, the numbers steadily increasing until by 1964 there were 246 plants and 162 flowering spikes. This remarkable increase can only be attributed to the delicate and back-breaking task of hand-pollination, which Mr Wilks and his helpers undertook each year. In 1965 there were 205 flowering spikes and soon afterwards the task of hand-pollinating all the flowers had to be discontinued. In 1976 the drought resulted in a total lack of flowers, but in 1977 plants reappeared on the main site and, more exciting, appeared for the first time in cleared woodland a short distance away.

In 1958 seed from the East Kent colony was broadcast on another Kent reserve and in 1965 three plants flowered. Two more flowered in 1976, and it is now well established there.

The Monkey Orchids in the original Chilterns site continued to hang on, two small delicate plants flowering since 1958, one bearing ten flowers. By 1967 there were eight flowering spikes, some of the plants having the healthy stature of those in Kent. In the previous year Monkey Orchids had been found in a second site in the Chilterns. It occurred for the first time in south-east Yorkshire in 1974, the colony increasing to twenty-five individuals and nine flowering spikes before it was destroyed in 1981.

The Monkey Orchid has two egg-shaped tubers and a few fleshy roots, the flowering stem measuring 10 to 30 cm in height. There are three to four blunt glossy basal leaves which are often keeled. Above these are two to three sheathing stem leaves. The mauvish bracts are very short.

The flowering spike of the Monkey Orchid is roughly globular in shape, the flowers rather densely packed, and the whole spike looking oddly untidy. This is due partly to the long lobes of the labellum, and partly to the unique habit of the Monkey Orchid in having flowers at the top of the spike opening first—a complete reversal of the pattern in all other orchid species.

The sepals and two upper petals form a long, pale, pointed hood, similar to that of the Military Orchid. The colour is whitish, tinged with lilac or rose. It differs from the Military Orchid in that the segments comprising the hood are looser, slightly spreading, and marked with a series of dark-mauve dots and streaks, which do not form the prominent lines present inside the hood of the Military Orchid.

The labellum bears four lobes and a median tooth like the Military Orchid, but all the lobes are long and narrow, resembling the arms and legs of a dancing monkey. Even the median tooth is longer and narrower, forming the monkey's tail. The tips of the lobes are usually deep pinkish mauve, and tend to curl forwards. The base of the labellum is paler and bears a small triangular patch of bright mauve spots formed by small tufts of brightly coloured hairs. The spur is whitish pink, and about half the length of the ovary.

The Monkey Orchid has a faint vanilla scent, but it is thought that the spur does not contain nectar. The flowers are visited by flies, bees, and butterflies. M. J. Godfery recorded one butterfly with seven pollinia attached to it. Judging by the records from East Kent, the Monkey Orchid can flower seven years after seed is set.

The Monkey Orchid only grows on calcareous soils which have good drainage and, although it appears to like moderate shade and shelter, it flowers equally well in woodland and in open rough downland. There it is accompanied by Fragrant Orchid, Lizard Orchid, Man Orchid, Pyramidal Orchid, and the usual flora of rough downland such as Salad Burnet, Felwort (*Gentianella amarella*), and Dropwort (*Filipendula vulgaris*). The flowers are at their best in late May and early June.

There is considerable variation both in the stature of the plants and in the colour of the labellum. The Kentish plants tend on the whole to be more robust with a strongly coloured and spotted labellum, the plants in the original Chilterns site being of small stature, very pale, and with a scarcely marked labellum.

The hybrid, *Orchis militaris* × *simia*, occurred in the nineteenth century in the Thames Valley area, when both species were much commoner than they are now. In May 1985 the hybrid *Aceras anthropophorum* × *Orchis simia* flowered in Kent.

Listed in the *Red Data Book*.★

38 Burnt Orchid

Orchis ustulata L.

The Burnt Orchid is very like a tiny edition of the Lady Orchid and, although it is becoming much rarer, as old pastures are disturbed, this charming little orchid can still be found in large numbers on some old chalk downland.

There are two globular tubers, the stem rarely exceeding 15 cm in height, most of the plants being only 6 to 7 cm high. At the base of the stem are two to five broad and channelled leaves, 2 to 3 cm long, with well-marked parallel veins. Above these are several stem leaves which sheath the stem. The bracts are pointed, reddish brown, and about half the length of the ovary.

The flower spike is dense and cylindrical. The unopened flower buds are a rich red-brown colour, so that the top of the flower spike looks as if it had been burned, hence the name Burnt Orchid. The neat hood is formed by the sepals and smaller narrow upper petals, the red-brown colour tending to fade as the flowers mature.

The labellum is white, contrasting beautifully with the dark hood, and marked with a series of bold crimson spots. The labellum has two rounded lateral lobes and a longer central lobe which is shallowly lobed at the tip but, unlike the Lady Orchid, has no median tooth. The red spotting extends on to all parts of the labellum. At the base of the labellum is a short down-curving conical spur, about a quarter the length of the ovary.

The flowers of the Burnt Orchid have a strong, sweet scent which is attractive to insects. M. J. Godfery noted a large fly, *Echinomyia magnicornis*, visiting the flowers and removing the pollinia. Seed is set in a high proportion of the flowers.

The Burnt Orchid has one of the most protracted growth periods of any of the British Orchids. After seed has been dispersed the protocorm grows underground for ten to fifteen years before the first aerial leafy stem is produced, and several further years elapse before there is sufficient tuber growth for the plant to be able to flower. It follows that such a species can only survive where the grassland is never ploughed.

Since the Burnt Orchid is a plant of old, undisturbed chalk pastures, one main area of distribution is in south and south-east England. While it is without doubt declining in status, both as regards stations and total numbers, it still occurs in East and North Kent, Sussex, Hampshire, Berkshire, Wiltshire, Gloucestershire, and north Bedfordshire. On the downland there it may sometimes flower in considerable numbers, in sheltered sunny places which are always well drained and often south facing. Being so small the plants cannot flourish where the grass is long. It is accompanied by Salad Burnet, Horseshoe Vetch (*Hippocrepis comosa*), Green-winged Orchid, and especially the Chalk Milkwort, often favouring areas of Juniper scrub (*Juniperus communis*). Although the Chalk Milkwort flowers are ice blue in colour, they are the same size and shape as the Burnt Orchid flowers and in bright sunshine it is remarkably difficult to distinguish the orchid spikes among them.

The Burnt Orchid has a second main area of distribution on the limestones of the Midlands and north of England, in North Lincolnshire, Derbyshire, Westmorland, Cumberland, and Durham, but only 15 sites remain. It is absent from Wales, Scotland, and Ireland.

The flowering period in the south of England is from early May to mid-June, although in the north it may be a little later. The numbers flowering vary greatly from year to year, depending on the weather conditions. A yearly census of one site in Sussex gave the following figures. Note the nil count for the drought of 1976.

1966	178	1978	137
1967	88	1979	134
1968	484	1980	26
1969	649	1981	47
1970	10	1982	5
1971	318	1983	234
1972	773	1984	nil (cold)
1973	1858	1985	332
1974	262	1986	469
1975	415	1987	173 (cold)
1976	nil (drought)	1988	2162
1977	91		

Careful marking of individual plants has shown that they may flower in at least two consecutive years. The flowers wither very quickly, and once the seed has been set it is extremely difficult to find the plants, even when one has marked the site, and certainly by mid-June in the south of England all trace of them has gone.

During the last six years Burnt Orchids have been found in five sites in East Sussex and one site in Hampshire, coming into flower in July. The spikes appear in bud during the first week of July and the flowers persist to the end of the month. This late-flowering form has taller and less dense flowering spikes, although this may merely be the result of growing in grass which is taller at that time of the year. The individual flowers are smaller and darker, with a narrower labellum. The Sussex sites are quite separate from other places in the county where the early form grows and are of both north- and south-facing aspect. The two flowering periods are quite distinct with a gap of nearly a month between them, and it will be interesting to discover if this pattern is repeated in other areas and whether we are seeing the emergence of a late-flowering variety.

Plants with all white flowers, or with straw-coloured hoods and white labella, have occasionally been recorded. No hybrids are known in Great Britain.

39 Green-winged Orchid

Orchis morio L.

Only a few years ago the Green-winged Orchid was widespread, especially in the south of England, and not uncommon, but the drainage and cultivation of the old damp pastures it favours had led to a drastic reduction in numbers. Although still locally abundant where it does occur, it must now be considered a threatened species.

There are two rounded tubers, from which the rather short stem arises. Although plants have been recorded as tall as 40 cm they are normally much shorter, in the range 6 to 15 cm. At the base of the stem there are as many as seven narrow, bluish-green leaves, with a further two or three smaller pointed leaves sheathing the stem. The leaves of the Green-winged Orchid are never spotted, as distinct from the purple-spotted leaves of the related Early-purple Orchid. The bracts are fairly broad, pointed, and about two-thirds the length of the arched ovary. In purple-coloured flowers they are usually veined and edged with mauve.

The flower spike is relatively short and composed of six to twelve well-spaced flowers. There is a considerable variation in colour within any population, most of the plants being purple, although the colour can range from blackish purple through every shade to pale lilac. Flowers will also be found in varying shades of pink and a small percentage of pure-white flowers will be found in most large colonies.

The sepals and two narrow upper petals form a loose and rounded hood, as distinct from the spreading perianth segments of the Early-purple Orchid. The two lateral sepals are marked with six to seven bold parallel lines which are green or dark bronze coloured. These are especially prominent in albino flowers, and from them the species gets its name. The albino plants are of normal robust growth and observation over the last ten years in one colony in Sussex has shown that they occur regularly as about 1 per cent of the population.

The labellum is broad and three lobed, the two lateral lobes more or less reflexed, and the edges of all the lobes crenated. The central lobe is longer and broader, the base being paler or almost white, and marked with a group of purple dots. At the base of the labellum is the broad spur, as long as the ovary and usually pointing slightly upwards. The end of the spur is blunt and swollen, and may be notched.

The flowers have a powerful sweet scent which is attractive to insects, especially bees. There is no free nectar in the spur, but sugar is present in the spur wall. The bursicles are mauve and enclose pale yellow pollinia, a bee having been recorded with sixteen of these pollinia stuck to its head. The pollinia have separate viscidia, and after removal swivel downwards and forwards with considerable rapidity, the movement being completed within thirty seconds. Seed is set in the majority of capsules and, since most plants only flower once, this is the chief way in which the species can persist and spread. The gathering of the flowers of the Green-winged Orchid has certainly contributed to the decline in its numbers and underlines the rule that the flowers of wild orchids should never be picked if we are to continue to have the pleasure of finding them.

Development from seed is very rapid. The first leaf appears in the spring of the second year, and the first tuber in the year following. Flowers are produced after a further few years' growth, after which the plant usually dies. It is virtually impossible to check this with large colonies but it certainly seems true of small groups of plants where one can remember the positions and colours of the various flowers.

The Green-winged Orchid is most densely distributed in the southern and eastern part of England. In Wales it is commoner in the coastal areas, while in Ireland it is found right across the central plain. The first Scottish record was a single plant in Banffshire in 1958. In 1981 a huge colony was found on the Ayrshire coast.

The Green-winged Orchid is always local in occurrence, favouring old meadows and pastures on calcareous soils and also on clay soils below the downs, while in North and South Wales it flourishes in the slacks of sand-dunes. In many areas it appears in churchyards and on lawns, but never in woodland. The ground is always rather damp, and often has a thriving population of Cowslips (*Primula veris*) and the little Adder's tongue. One thriving colony, which now numbers over 1,500 spikes, only appeared seventeen years ago when a small area of oak- and Hazelwood was cut down. The area was seeded with grass and soon afterwards the first orchids appeared. With them are plants of Common Twayblade and Common Spotted-orchid which have obviously come from more extensive colonies in the remaining wood, and once again the Adder's-tongue thrives in the turf alongside the orchids.

The flowering period is from late April to early June, the flowers usually being at their best in mid-May, and there is no more lovely sight than a field purple with thousands of Green-winged Orchids.

Variations in colour from shades of purple to pink or white have been widely reported and are not uncommon.

There is no mention of hybrids by Dr C. A. Stace in *Hybridization and the Flora of the British Isles*, but hybrids with *Orchis mascula* are mentioned by Summerhayes (1951) and described as being intermediate in character. It occurred in Westmorland in 1985.

40 Early-purple Orchid

Orchis mascula L.

The Early-purple Orchid is one of our commonest orchids and has a wide distribution on base-rich soils throughout the British Isles. It has been known and described for many years, and, as Summerhayes (1951) pointed out, is referred to by the Queen in *Hamlet* under the name Long Purples. The local name of Rams-horns, which Summerhayes noted for Sussex, is still in current use as a popular name for them.

The Early-purple Orchid has two ovoid and bloated tubers, from whose appearance the scientific name derives, and from these arises the stem 15 to 60 cm high. At the base of the stem is a sizeable rosette of four to eight blunt leaves, which are shiny on the upper surface and usually heavily marked with irregular purple blotches. There are two to three stem leaves, which sheath the stem and may also bear a few small spots. The degree of spotting varies greatly, from plants with unmarked leaves to those which are heavily blotched. The bracts are fairly long, equalling the ovary.

The flower spike is lax and bears on average some twenty to fifty reddish-purple flowers, far more than the Green-winged Orchid. Woodland plants are usually taller with well-spaced flowers, while those of open downland are squatter and have denser spikes.

All the floral parts are purple. The two lateral sepals are spreading at first, but in the well-opened flower they are folded right back like a pair of wings, until they almost touch each other behind the flower. The upper sepal and two upper petals form a loose hood.

The labellum is large and three lobed, the central lobe longest, notched at the tip, and the two lateral lobes more or less reflexed. The separation between the lobes is very variable, ranging from labella which are deeply cleft between the lobes, to labella which are entire and almost flat. The edges of all the lobes are crenated. The centre of the labellum is paler or yellowish, marked with a few dark spots, but never showing the contrast which is so marked in the Green-winged Orchid. The spur is long, stout, and blunt, curving or pointing upwards and equalling the ovary.

The scent of the Early-purple Orchid is strong, reminding most people of tom-cats. Flowers with a strong scent of honey or vanilla have been reported from places as far apart as Ardnamurchan in Argyll, Ashdown Forest in Sussex, and Sweden. It is possible that the foul scent develops after pollination has occurred.

The flowers are visited and pollinated by bees (the mechanism is described in detail on pages 10–11). Seed is set in a high proportion of flowers. It develops more slowly than the Green-winged Orchid, forming the first tuber in the second year and the first leaf in the fourth. The plant then matures slowly until a full rosette is formed and a flower spike produced, after which it usually dies. The species is thus dependent on seed production, and this gives rise to big yearly fluctuations in the number of flowering plants. The species can also persist in the vegetative state in dark woodlands for many years, appearing as a mass of blooms when the trees are felled.

The Early-purple Orchid occurs in nearly every vice-county in Great Britain, including the Orkneys and Shetland, but never in very acid or very wet situations. In the south of England it occurs in woodlands with Bluebells (*Endymion non-scriptus*), Dog's Mercury, Common Twayblade, Greater and Lesser Butterfly-orchids, Lady Orchid, and Common Spotted-orchid; it also occurs on the open chalk downland where the plants are shorter and stouter, often growing with Cowslips, and frequently on banks and road verges. It is especially lush and abundant in the grassy walls and hedgerows of Devon and Cornwall, and grows with sheets of Bluebells on the hillsides of the north of England, Scotland, and the Isle of Man. It is frequent along grassy cliff tops in the west and south-west of England, and equally frequent in the limestone country of the Burren in west Ireland.

The Early-purple Orchid lives up to its name and is often the first orchid to flower in the spring. I have even seen flowers fully out on 13 April 1966, in Sussex, standing up bravely above a heavy fall of snow. The flowering period is long and plants will still be found in flower in mid-June.

Most flowers are reddish purple, but pale pink specimens are not uncommon, predominating in one colony high up in the Dales of Yorkshire. Albino flowers do occur, but in this species they do not seem to be particularly common. C. B. Tahourdin in 1924 recorded a plant near Ballyvaughan in Co. Clare which had all white flowers and a purple-spotted labellum. Plants with pale flowers covered with fine

< 32 >

< 33 >

<
>
34
>

<
>
35
>

36

<
37
>
∨

<
38
>

39

39

40

albino form

ssp. *okellyi*

ssp. *hebridensis*

< **41** >

< **42** >

 43 ∧ ∧

ssp. *pulchella*

ssp. *coccinea*

ssp. *coccinea*

< **43** >

ssp. *cruenta*

ssp. *ochroleuca*

< **43** >

ssp. *cruenta*

ssp. *ochroleuca*

< **43** >

< **44** >

∧ **45** < >

ssp. *majaliformis*

ssp. *occidentalis*

∧
46
<
∨

ssp. *occidentalis*

ssp. *cambrensis*

ssp. *cambrensis*

∧
46
>
ssp. *scotica*

<
47
>

< **47a** >

< **48** >

49

albino fo

mauve flecks were recorded by D. Pearson from Cleeve Hill, Gloucestershire, in 1988.

One abnormal plant was found in 1967 near the foot of Beachy Head in East Sussex. The basal rosette consisted of seventeen massive leaves, with a fasciated stem 40 cm high.

41 Common Spotted-orchid

Dactylorhiza fuchsii (Druce) Soó

[*Orchis fuchsii* Druce; *Orchis maculata* auct. *meyeri* Reichenbach; incl. *Orchis maculata* ssp. *okellyi* Druce]

The two spotted-orchids and five marsh-orchids comprise the group known as the Dactylorchids, with the generic name *Dactylorhiza*, literally 'hand rooted', on account of the palmately lobed tubers. The inexperienced botanist may have some difficulty in identifying them, but with the photographs (Pls. 41–7) as a guide and plenty of field experience it should not prove too difficult. The various species have probably only become differentiated in relatively recent times and this close relationship results in plants which cannot be easily defined. It also permits wholesale hybridization within the genus, with the production of hybrid swarms further to confuse the unwary.

The Common Spotted-orchid is widely distributed throughout the British Isles, being commonest in the south and east of England where the definitive form occurs. The tubers are divided into four or five tapering lobes, above which are a few fleshy roots. The stem is 15 to 45 cm high, with as many as twenty leaves at its base. The lowest leaf is the broadest, rounded at the tip, and shorter than the leaf above it. There are also three to five clasping stem leaves which grade into the lower bracts. The leaves are usually spotted, the spots being formed by transversely elongated blotches. There is great individual variation in spotting, some plants having few if any marks on the leaves, while in others the basic green colour is almost obliterated by massive brownish-purple blotches. The bracts are long and pointed, exceeding the ovary.

The flower spike is long and tapering, with many densely packed, pale lilac flowers. The lateral sepals are spreading, marked with darker lines and dots, the upper sepals and two upper petals forming a loose hood.

The labellum is three lobed, the lateral lobes rhomboidal and the longer central lobe triangular. The labellum is marked by a prominent symmetrical double loop of broken lines and dots in darker mauve, and at its base is a slender, straight, cylindrical spur, just over half the length of the ovary.

The flowers are faintly scented and are visited by bees of many species as well as syrphid flies, which feed on the sugar secreted in the wall of the spur. The two pollinia are carried on separate viscidia. Pollination is efficient and most capsules bear seed. The first leaf is developed in the second year and the number of leaves increases year by year until the mature plant flowers. It then appears to be perennial for some years, the flower spikes increasing in stature, while vegetative multiplication frequently leads to the formation of clumps of flowering plants. Plants can also remain in a vegetative state for many years if conditions become temporarily unfavourable.

The Common Spotted-orchid is widely distributed and flourishes in a variety of habitats, showing some preference for calcareous soils. It grows with equal facility in woodland, open downs, marshes, and dunes, particularly favouring railway cuttings and old chalk-pits where it is more sheltered. There it is accompanied by the Common Tway-blade, Fragrant Orchid, Greater and Lesser Butterfly-orchids, Bee Orchid, Fly Orchid, Pyramidal Orchid, and the Marsh-orchids, especially the Southern Marsh-orchid. In one interesting site near Sway in the New Forest the Common Spotted-orchid flourishes in a railway cutting where the underlying calcareous soil is exposed, while the Heath Spotted-orchid flowers barely one and a half metres away on the undisturbed acid heath. The flowering period is long, from mid-May in sunny sites to the end of July.

The Common Spotted-orchid also occurs in two quite distinct sub-species. *Dactylorhiza fuchsii* ssp. *hebridensis* (Wilmott) Soó is the dominant form in the Hebrides, where it occurs in vast numbers in the machair of the Outer Isles, and on Tiree, Jura, and Islay. It is a dwarf plant about 10 cm high, with five to six rather narrow leaves, which are finely spotted or unmarked. The flower spike is short, dense, and almost spherical in appearance, the basic colour being a deep rosy purple. The flowers have a strongly three-lobed labellum, the central lobe longest, marked heavily with dots and streaks which do not form well-defined loops. On the machair it forms sheets of pinkish mauve, with Eyebright (*Euphrasia* spp.), White Clover, Wild Thyme (*Thymus drucei*), and Frog Orchids. This sub-species has also been recorded in East Cornwall, West Sutherland, South Kerry, West Galway, and West Donegal.

The second sub-species, *Dactylorhiza fuchsii* ssp. *okellyi* (Druce) Soó, was first described by P. B. O'Kelly at Ballyvaughan in Clare. The leaves are slender and normally unspotted, the lowest leaf being long and narrow although the tip is always rounded. The spike is square topped, and the strongly scented flowers are small and white, or just faintly marked. The labellum is sharply divided into three almost equal lobes. This form is common in north and west Ireland, in Down, Fermanagh, Leitrim, and the Burren of Clare and Galway. It has also been recorded in parts of north-west Scotland, including Sutherland, while I found it in 1973 in a small calcareous marsh on the south side of Tiree in the Hebrides. It has also been found in the Isle of Man.

Care must be exercised in differentiating ssp. *okellyi* from true albino plants of the normal form of *D. fuchsii*. These have stouter leaves and large, only faintly scented flowers in a conical spike. The labellum is the same shape as that of the normally coloured *D. fuchsii*. Flowers which appear to be albino often prove, on close inspection, to have faintly marked labella, albinism not being very common in this species.

Abnormal flowers have been recorded, a spike with three terminal heads being found near Warlingham in Surrey in 1920, while plants with inverted flowers are not uncommon. In a bog near Sheringham in Norfolk a number of the spikes consisted of bracts only, set with the odd flower at intervals of several centimetres.

The Common Spotted-orchid hybridizes with many other orchids. These hybrids, which are described later, involve the following species— Frog Orchid, Fragrant Orchid, Heath Spotted-orchid, Early Marsh-orchid, Southern Marsh-orchid, Northern Marsh-orchid, Broad-leaved Marsh-orchid, and Irish Marsh-orchid.

42 Heath Spotted-orchid

Dactylorhiza maculata (L.) Soó ssp. *ericetorum* (E. F. Linton) Hunt & Summerhayes

[*Orchis ericetorum* (E. F. Linton) A. Bennett; *Orchis elodes* Grisebach; *Orchis maculata* L.]

The Heath Spotted-orchid is the complete antithesis of the Common Spotted-orchid, being rare on calcareous soils, never growing as a truly woodland plant, but flourishing on moors and acid heaths. Because of this, although it is widespread throughout Great Britain, it is commonest in the northern and western areas which have the greatest stretches of acid peaty soil.

The tubers and roots are similar to those of the Common Spotted-orchid although the lobes of the palmate tubers are very long and tapering. Most plants have a rather slender, ridged stem 10 to 25 cm high, but occasionally massive plants over 40 cm high may be found in moist, sheltered localities, with proportionally thicker stems.

The four to eight leaves are all narrow and pointed, the lowest leaf never being short or rounded as in the Common Spotted-orchid, while the stem leaves are narrow, pointed, and clasping. The spotting on the leaves is seldom heavy, and may be very faint or even absent, the shape of the spots being more or less circular. The bracts exceed the ovaries in length.

The flower spike is short and pyramidal, with an over-all colour of a very pale pinkish white. The lateral sepals are spreading, rather short and blunt and marked with faint lines and dots. The upper sepal and the two upper petals form a loose hood.

The labellum is very broad and flat, the large lateral lobes being rounded, while the median lobe is small, triangular, and often shorter than the lateral lobes. The markings on the labellum consist of numerous small reddish dots and short lines which extend over most of the surface and do not form the pronounced double loop which is such a feature of the Common Spotted-orchid. The broad frilled labellum, with its delicate markings, makes the Heath Spotted-orchid a most attractive plant. The spur is a little shorter than the ovary, and much more slender than the spur of the Common Spotted-orchid.

The flowers have a very faint scent and are visited by bees and flies.

Seed production is good and the plants develop in a similar manner to those of the Common Spotted-orchid. Like them they are perennial for some years on reaching maturity, but show less tendency to multiply vegetatively and form clumps, probably because of the relative poverty of the soil in which they grow.

The Heath Spotted-orchid is extremely common on moorland and acid heath throughout Great Britain, in the New Forest, Dartmoor, Exmoor, Cornwall, throughout Wales, and on moorland from north England throughout Scotland, the Orkneys, Shetland, all the Hebrides, and the entire north and west of Ireland. As would be expected from its soil preferences, it is much rarer in south-east England and the Midlands.

It grows in profusion on dry heaths and moors with Cross-leaved Heath (*Erica tetralix*) and Purple Moor-grass, often being the only orchid to be found there, and it flourishes equally in slightly wetter areas with some *Sphagnum* moss, Cottongrass (*Eriophorum* ssp.) Lesser Butterfly-orchid, and Common Butterwort (*Pinguicula vulgaris*).

In the south of England it tends to flower a little later than the Common Spotted-orchid, from mid-June to late July, while in Scotland and the Hebrides it may well flower earlier. In 1973 on Tiree it was already over in early July by the time the Hebridean orchid *Dactylorhiza fuchsii* ssp. *hebridensis* was fully out.

Albinism appears to be more common than in the Common Spotted-orchid, and I have found a high proportion of pure-white flowers in colonies near Loch Maree and Loch Lomond in Scotland, and again in the New Forest area of Hampshire. In the latter area there is considerable variation in colour from white, through all shades of lilac and pink, to plants with bright pink flowers heavily marked with dark magenta. Morphological variation is not rare, and in 1964 near Sway in the New Forest I found a plant with inverted flowers and no ovaries.

The Heath Spotted-orchid hybridizes freely, and hybrids with the following species have all been recorded—Fragrant Orchid, Common Spotted-orchid, Early Marsh-orchid, Southern Marsh-orchid, Northern Marsh-orchid, Broad-leaved Marsh-orchid, and Irish Marsh-orchid.

The Marsh-orchids

While the identification of the two spotted-orchids is a relatively simple matter, the identification and even the definition of the various marsh-orchids species, as encountered in the field, is extremely difficult.

The whole group is obviously still evolving rapidly, and species hybridize freely among themselves, and with the spotted-orchids, so that hybrid swarms are encountered as frequently as plants of apparently unmixed parentage. Once this fact has been accepted, the species must be defined, and all other forms consigned to these as sub-species and varieties, or banished to the realm of the hybrids.

Since the first edition of this book in 1980, a considerable body of research work has accumulated on marsh-orchids, which necessitates substantial rethinking of their nomenclature. One species, *Dactylorhiza lapponica*, has been described new to the British Isles. Further reorganization of the group is inevitable as new light is thrown on their complexities, but the list of marsh-orchids on p. 125, with their various sub-species, defines the present state of play.

43 Early Marsh-orchid

Dactylorhiza incarnata (L.) Soó

[*Orchis latifolia* (L.) Pugsley; *Orchis incarnata* L.; *Orchis strictifolia* Opiz; incl. ssp. *cruenta* (*Müller*) *Vermeulen*]

Despite the impressive range of colours which the Early Marsh-orchid demonstrates, the species is easily identified by the shape of the leaves and flowers.

The palmate tubers are divided into two to four tapering lobes, above which are a number of long spreading roots. The stem is fairly stout but hollow and 10 to 30 cm in height, most of the plants being on the short side.

At the base of the stem are four to seven erect, sword-shaped yellow-green leaves. They are keeled, with hooded tips, and in every form, except ssp. *cruenta*, they are unspotted. The upper leaves sheath the stem and grade into the bracts, which are long, projecting well beyond the flowers and often marked with reddish purple.

The Marsh-orchids

Early Marsh-orchid *Dactylorhiza incarnata*	ssp. *incarnata*	widely distributed
	ssp. *pulchella*	widely distributed
	ssp. *coccinea*	mainly sand-dunes and machair
	ssp. *ochroleuca*	East Anglian fens
	ssp. *cruenta*	western Ireland and Wester Ross
Southern Marsh-orchid *Dactylorhiza praetermissa*	ssp. *praetermissa*	south Britain
	ssp. *junialis*	south Britain
Northern Marsh-orchid *Dactylorhiza purpurella*	ssp. *purpurella*	north and north-west Britain and Ireland
	ssp. *majaliformis*	north and north-west Scotland
Broad-leaved Marsh-orchid *Dactylorhiza majalis*	ssp. *occidentalis*	western Ireland
	ssp. *cambrensis*	Wales, including Anglesey
	ssp. *scotica*	North Uist only
Irish Marsh-orchid *Dactylorhiza traunsteineri*	no sub-species	widely distributed
Lapland Marsh-orchid *Dactylorhiza lapponica*	no sub-species	north-west Scotland

The flower spike is dense, bearing twenty to thirty rather small flowers which have a curious pale flesh-pink colour, from which the species derives its name. The lateral sepals are folded right back so that they stand erect above the loose hood formed by the upper sepal and the two upper petals. The lateral sepals bear small spots and loops of darker red.

The labellum is shallowly three lobed with the sides strongly reflexed, so that it appears very narrow and slightly convex. It bears a double loop of dark red, which encloses a central area marked with fine dots and lines. This red double loop is a prominent feature in all the colour forms, except ssp. *ochroleuca*. The spur is short, stout, and conical, and is another feature which appears to be dominant in hybrids with the spotted-orchids.

Seed is set in a high proportion of flowers and, judging by the numbers of hybrids produced, there must be considerable insect activity involved, probably by bees. Crab spiders are not uncommonly found in the flower spikes, as they are in the Fragrant Orchid. The first leaf is produced in the second year and the first tuber in the fourth, after which several seasons elapse before the plants are mature enough to flower. They then persist and flower for some years.

The Early Marsh-orchid is widely distributed throughout the British Isles in water-meadows, damp hayfields, bogs, marshes, fens, and dunes, preferring a soil with an alkaline or neutral pH. It seldom occurs in large numbers and is decreasing rapidly in those areas of the country where there has been extensive draining of wetlands.

The flowering period is from late May throughout June. In mixed populations of the five subspecies, the difference in flowering time (as much as fifteen days) may act as a reproductively isolating factor ensuring the continuity of the subspecies.

While the normal form of the Early Marsh-orchid favours alkaline or neutral soils, the next colour form, ssp. *pulchella* (Druce) Soó, flourishes in rather more acid areas. It will be found in marsh or fen sites dominated by *Schoenus nigricans* with a pH 6.2–7.2, in poor fen with no *Schoenus* or *Sphagnum* moss, and in acid bogs carpeted with *Sphagnum* and having a pH 5–6. In morphology it is identical with the normal form, but the colour is a mauve purple, bearing the usual red loops and dots on the labellum. The two colour forms are not exclusive and will be found side by side, but in certain hilly and acid areas, such as occur in North Wales and the New Forest of Hampshire, this form predominates. It is also recorded from many areas of mainland Scotland, the Hebrides, Wales, in England west to Cornwall, and in Ireland. It would appear that the *D. incarnata* var. *cambrica* Pugsley from Borth Bog in Dyfed in 1935 and the *Orchis traunsteineri* of Druce from the neighbouring Tregaron Bog were both this colour form of *D. incarnata*.

The third colour form, ssp. *coccinea* (Pugsley) Soó, is the most striking of all the marsh-orchids, being bright vermilion, especially when newly opened. The plants tend to be shorter and stouter than the normal form, with four to seven thick, dark-green, strongly keeled and hooded leaves. They look rather like little fat hyacinths, and the sight of thousands in flower in the dune slacks they favour is truly astonishing.

Ssp. *coccinea* grows in moist dune slacks where the sandy humus overlying pure sand may have a *p*H 6.2–7.6. It also grows in stabilized dune grassland. In Limerick Heslop-Harrison recorded it with ssp. *pulchella*, the latter growing on wet lake margins, while ssp. *coccinea* occupied the drier ridges between the hollows. It grows in similar mixed populations in drier pastures on the limestone plain of central Ireland. In addition to these areas in Ireland it has been recorded from Clare, Dublin, and Donegal. It is widely distributed in Norfolk, North and South Wales, the Isle of Man, West Sutherland, and the Hebrides. Associated species include Creeping Willow, Burnet Rose, Marsh Helleborine, and in some areas the Green-winged Orchid and the dune form of the Fen Orchid. Where ssp. *coccinea* occurs with the normal form in North Wales it tends to flower ten to fourteen days later.

The fourth colour form, ssp. *ochroleuca* (Boll) Hunt and Summerhayes, was first found in Britain in a Norfolk fen by J. E. Lousley in 1936, and confirmed by Pugsley, who later found it in another West Norfolk fen, the only form present and characterized by its tall growth and straw-coloured, unmarked flowers. These have a broad, strongly three-lobed labellum and a large spur. The yellow colour is unique among marsh-orchids. This sub-species is now much reduced in its East Anglian sites. The peak flowering period is usually two weeks later than that of the normal form. Ssp. *ochroleuca* is abundant in many fen sites on the island of Gotland in the Baltic, growing with ssp. *pulchella* and ssp. *cruenta* and showing no tendency to intergrade.

The fifth form, ssp. *cruenta* (O. F. Müller) P. D. Sell, was first recognized in the British Isles by J. Heslop-Harrison in 1949 in west Ireland. It is a fairly small plant 10 to 30 cm high, with four to six leaves. The distinguishing feature is that the leaves are spotted with reddish purple on both surfaces, the spots being irregular and frequently elongated parallel to the leaf edge.

The spike is dense, bearing about fifteen lilac-purple flowers, the bracts also being tinged with purple. The labellum is entire or slightly lobed, the sides of the labellum not much reflexed, the markings being indistinct and often lacking the prominent double loop.

It is widespread but decreasing in the area of the lakes south-west of Mullaghmore between Lough Bunny and Corofin in Co. Clare, where it flowers in highly calcareous lakeside fens accompanied by *Schoenus nigricans*, Water Germander (*Teucrium scordium*), and Fen Violet (*Viola stagnina*), with Shrubby Cinquefoil on slightly drier ground. It is also recorded in north-east Galway, East Mayo, and was discovered for the first time on mainland Britain in West Ross in 1982. It is said to be late flowering at the end of June and during July, but M. C. F. Proctor records it in the Burren fully in flower at the end of May.

The evolution of ssp. *cruenta* is the subject of debate.

Hybrids are recorded between the various subspecies of the Early Marsh-orchid and the following species—Common Spotted-orchid, Heath Spotted-orchid, Southern Marsh-orchid, Northern Marsh-orchid, Broad-leaved Marsh-orchid, Irish Marsh-orchid, and Fragrant Orchid.

44 Southern Marsh-orchid

Dactylorhiza praetermissa (Druce) Soó

[*Orchis praetermissa* Druce; incl. *Orchis pardalina* Pugsley and *Dactylorhiza praetermissa* ssp. *junialis* (Vermeulen) Soó]

The Southern Marsh-orchid is well distributed throughout the south of England as far north as Durham, but is absent from Scotland and Ireland where it is replaced by the Northern Marsh-orchid. In a few sites in Lancashire, Derbyshire, and Wales the two species grow together, and there hybrids between the two have been recorded.

The Southern Marsh-orchid is a robust plant. Arising from the large palmate tubers, the stem which is usually hollow and soft may reach 70 cm in height.

The yellow-green leaves are unspotted, and as many as nine form a bushy cluster around the base of the stem. They are broad and flat, and are not hooded at the tips. The upper stem leaves are sheathing and merge into the bracts, which are often coloured and project beyond the flowers.

The flowering spike can be massive, and I have seen several plants in damp dunes in North Wales with spikes 20 cm in length, containing over a hundred flowers. The individual flowers are larger than those of the Early Marsh-orchid. The lateral sepals initially are spreading, but later become more erect. The other perianth segments form a loose hood.

The labellum is broad and even concave, the edge in many cases being entire, so that it appears spoon shaped. At the centre of the labellum is a paler area which is marked with very small dots, but never with the double loop so typical of the Early Marsh-orchid. The spur is shorter than the ovary, thick and blunt, and may be slightly down-curved.

The typical Southern Marsh-orchid is a pale lilac mauve, but in many areas of south and south-west England plants will be found with dark-magenta flowers. The spots on the labella of these darker flowers are often well marked, but this seems to be normal variation within the species, and unless other morphological differences are present is not indicative of hybridization.

The Leopard Marsh-orchid (*Dactylorhiza praetermissa* ssp. *junialis*) is for the present best grouped with the Southern Marsh-orchid, although this arrangement may need to be altered if it is given specific status. The sub-specific name *junialis* has preference over the well-known name *pardalina*. In general shape and colour it much resembles the Southern Marsh-orchid, but the leaves are marked with dark spots and rings, from which it gets its name. Some authorities consider ssp. *junialis* to be a hybrid between the Southern Marsh-orchid and the Common Spotted-orchid, but there are a number of features which argue against this.

The hybrid is more robust with narrower, pointed leaves, which are spotted but do not bear rings. The bracts are shorter, and the flowers darker coloured with the looped markings on the labellum similar to those of *D. fuchsii*. In both the hybrid and ssp. *junialis* the spur is stout, but since this feature is present in marsh-orchid species and hybrids alike, it cannot serve to differentiate them. The fertility of the hybrid is not high.

Ssp. *junialis* always occurs with or near *D. praetermissa*, but in some areas of Cornwall where I have found it plentiful, *D. fuchsii* does not occur for a considerable distance around. In these colonies ssp. *junialis* often predominates and shows a greater degree of homogeneity than one would expect in a group of hybrids. Fertility seems to be high, judging by the number of ripe seed capsules formed.

Ssp. *junialis* has been recorded from Kent to Cornwall in south England and again in Suffolk. Dr Erich Nelson, in personal correspondence, remarks that ssp. *junialis* as found in the Isle of Wight, is midway in type between plants found in south England and those recorded in Holland. We should therefore keep an open mind on the Leopard Marsh-orchid until further evidence is presented upon which a decision can be made, and meanwhile view it as a sub-species of *D. praetermissa*.

The Southern Marsh-orchid is widespread throughout south England in fens, marshes, wet meadows, and dune slacks where the soil is base rich. Although the number of sites is decreasing it is still locally abundant, often accompanied by the Marsh Helleborine. It can also be found growing on dry chalk hills and especially in abandoned chalk quarries, with Common Spotted-orchid and Fragrant Orchid.

In Cornwall it can be found in good numbers alongside the margins of the streams as they flow out over the grassy tops of the sea cliffs.

The flowering period is from mid-June to the end of July.

Hybrids have been recorded with the following species—Fragrant Orchid, Common Spotted-orchid, Heath Spotted-orchid, Early Marsh-orchid, Northern Marsh-orchid, and Irish Marsh-orchid.

45 Northern Marsh-orchid

Dactylorhiza purpurella (T. & T. A. Stephenson) Soó

[*Orchis purpurella* T. & T. A. Stephenson]

The Northern Marsh-orchid was first recognized as a distinct species in 1920 by T. and T. A. Stephenson, its other common name being the Dwarf Purple Orchid.

It is a plant of moderate size, 10 to 25 cm high, the stem being rather slender with only a small cavity at its centre. The palmate tubers are divided into two to four lobes.

Ssp. *purpurella* is a plant of variable stature and form. Plants can be very robust, with broad, glaucous green leaves, but ssp. *purpurella* is tolerant of damp acid pastures, and there individuals may be smaller with narrower leaves. The leaves may be unspotted or bear a few spots near the leaf apex. The bracts are unspotted and not as long as those of the Early Marsh-orchid.

The shape of the flower spike is very distinctive and a useful guide to identification. It appears square-topped, as if someone had cut out the top third of the inflorescence with a pair of scissors. The colour is also distinctive, most plants being vivid purple and bearing fifteen to twenty flowers.

The lateral sepals are spotted and folded upwards, while the other perianth segments form a loose hood. The labellum is entire, or may show three shallow lobes, and is distinctively diamond shaped. The edges of the labellum are sometimes curled upwards when the flowers first open, but flatten out later. The labellum is heavily marked with dark lines and blotches, especially in the centre, but these never form loops as in the Early Marsh-orchid.

The spur is thick and conical, shorter than the ovary. Seed is set in a high proportion of flowers, and insects seem to be frequent visitors and pollinators, judging by the number of hybrids which occur with other species.

Ssp. *majaliformis* E. Nelson is a robust plant with a dumpy inflorescence. Ninety per cent of the plants have spotted leaves, the dull violet spots being larger than those of ssp. *purpurella*. The bracts are spotted also. The flowers are large and the labellum is broad, three lobed, bearing marks and broken lines. It is often intensely dark purple at the margins. The spur is wide and conical.

This sub-species has a very specialized ecological preference, being restricted to damp coastal localities within 100 m of the sea. The type plants were described from a site at Scrabster, Caithness, the main distribution being along the north and north-west coasts of Scotland.

Ssp. *purpurella* is widely distributed in Wales, Scotland, and Ireland. It has disappeared from the single site in the New Forest in Hampshire, but has recently been reported from Oxfordshire.

The Northern Marsh-orchid grows in bogs, marshes, and wet areas where there is some calcareous water and the soil is not over acid. There it is accompanied by sedges, reeds (*Juncus* spp.), Bog Asphodel, and in many places the Early Marsh-orchid. In both Wales and Scotland it grows in damp meadows up to 450 metres and along the margins of mountain roads where the ground is moist and the bracken is not entirely blanketing.

The main flowering period is from the second week of June, throughout July, but in some sites in the west of Scotland it has been recorded flowering in the last week in May.

Like all the Dactylorchids it hybridizes freely, and hybrids with two other genera have been recorded. The following species have been known to hybridize with Northern Marsh-orchid—Frog Orchid, Fragrant Orchid, Common Spotted-orchid, Heath Spotted-orchid, Early Marsh-orchid, Southern Marsh-orchid, and Broad-leaved Marsh-orchid.

46 Broad-leaved Marsh-orchid

Dactylorhiza majalis (Reichenbach) Hunt & Summerhayes

[incl. *Orchis occidentalis* (Pugsley) Wilmott]

The Broad-leaved Marsh-orchid was named from plants of the sub-species *occidentalis* which were first described in west Ireland. It is a difficult species to identify with accuracy as it is so variable in form.

The sub-species *majalis* is widely distributed in Europe. Recent work indicates that the taxon *Dactylorhiza majalis* ssp. *occidentalis* (Pugsley) Sell is entirely restricted to Ireland, growing in the west and south-west, with an outlying population on Aranmore Island, Donegal, recorded in 1981. Most of the records from north and north-west

Scotland should be referred to *Dactylorhiza purpurella* (T. and T. A. Stephenson) Soó ssp. *majaliformis* E. Nelson, *Dactylorhiza majalis* ssp. *scotica* E. Nelson, or *D. lapponica* (Laest. ex Hartman) Soó. Records from Orkney have still to be verified.

Dactylorhiza majalis ssp. *occidentalis* has palmate tubers and a stem 20 to 50 cm in height. There are about six broad glaucous basal leaves, the lower leaves being small while the central leaves curve away from the stem and are not stiff like those of the Northern Marsh-orchid. The spotting of the leaves is very variable, the brownish spots usually heavier than those in the Northern Marsh-orchid, while in some plants the leaves are unmarked. The upper part of the stem is angular, the bracts equalling the ovary and often suffused with purple.

The flower spike is dense, the flowers lilac purple and fairly large. The outer sepals are long and spreading, marked with darker rings and dots, while the inner segments are strap shaped with pointed ends which arch over the base of the labellum.

The labellum is broad, the central lobe more prominent, the edges of the lobes crinkled and slightly down turned. The labellum is marked with black spots and broken lines in a more or less symmetrical pattern, extending over the whole surface. The spur is shorter than the ovary and points downwards.

The Broad-leaved Marsh-orchid is usually a short, squat plant, but the characteristics of the leaves and the short, dense purple flower spike are the same as the Continental sub-species *majalis*. The important point is that it is a very early-flowering Marsh-orchid, coming into flower a good fortnight before the Early Marsh-orchid.

D. majalis ssp. *occidentalis* was first described in west and south-west Ireland, growing in damp boggy pastures and wet dune slacks, as well as in the limestone country of the Burren. There it is plentiful, being at its best about 20 May. R. Gorer recorded huge colonies between Ennis and Lisdoonvarna in Co. Clare, but suspected that there was some degree of hybridization with the Early Marsh-orchid.

Ssp. *cambrensis* R. H. Roberts has longer, narrower leaves, well pointed and spotted throughout their length. The flower spike is robust. The flowers are bright pinkish mauve, the central part of the broad, three-lobed labellum often being marked with darker broken lines. The spreading sepals are only lightly spotted, if at all. It has been recorded from Merioneth, Cardiganshire, Caernarvonshire, and Anglesey, growing both in marshes and in dune slacks.

Ssp. *scotica* E. Nelson is a dwarf plant 5 to 9 cm high with a few bright green, heavily marked leaves crowded at the base of the stem. The spots may merge into larger blotches. The short hollow stem bears a dense flower spike with bracts which may be spotted or suffused with purple. The flowers are violet purple with darker markings. The labellum is three lobed, 7–8 × 9–10 mm, with a spur 7–9 × 2.5–3 mm. Ssp. *scotica* grows in dune machair with a persistent high water-table— machair usually drains rapidly—and has only been recorded in North Uist. It flowers from early to mid May.

The documentation of putative hybrids of Broad-leaved Marsh-orchids needs extensive revision.

47 Irish Marsh-orchid

Dactylorhiza traunsteineri (Sauter) Soó

[*Orchis traunsteineroides* (Pugsley) Pugsley]

The Irish Marsh-orchid is a plant of very specialized habitat, being in the main restricted to calcareous fens, where it comes into flower two to three weeks earlier than any of the other marsh-orchids which grow there.

The tubers are lobed and the stem is 20 to 45 cm in height. The whole plant is of a slender and graceful habit, the stem soft and flexuous, curving between the stems of the reeds among which it grows. The three to five leaves are narrow and grasslike, the upper leaves being smaller, and in most cases they are pale green and unspotted.

The flower spike is lax, with ten to fifteen flowers which have a curious lilac tint quite unlike that of any other marsh-orchid. It is truly a case of 'once seen never forgotten'! The bracts are pink tinged, leafy, and pointed, projecting well beyond the flowers. The lateral sepals are very long, pointed, and spreading, while the upper sepal and two upper petals are nearly as long and arch forwards over the labellum.

The labellum is markedly three lobed, the central lobe being very long and pointed, and in some cases folded back beneath the labellum.

The surface is marked with darker spots and lines which do not form any particular pattern. Labellum shape and patterning are very variable. The spur is shorter than the ovary, broad and down pointing.

The Irish Marsh-orchid has a very scattered distribution and grows in a variety of habitats. In the calcareous fens of East Anglia in Norfolk, Suffolk, and Cambridgeshire, drainage and reclamation of the land have caused it to vanish from many previously flourishing sites. An additional and associated factor in this decrease is its ability to hybridize with other marsh-orchids. The hybrid with the Southern Marsh-orchid seems to flourish better than the Irish Marsh-orchid as the fens dry out, so that in one Suffolk fen since 1967 the hybrid has virtually taken over, and the Irish Marsh-orchid which once flowered there in profusion is now hard to find.

In south England it still grows in Berkshire and Hampshire, and was recently discovered in North Somerset. In Wales it has long been known from Anglesey, and has been recorded in the west Lleyn peninsula of Caernarvonshire from Cors Geirch to Ederyn. It is moderately widespread in Yorkshire, and has been found recently in Durham near Blackhall Rocks. Several populations have been found in wet hill flushes in north-west Scotland. In Ireland it is widely distributed right across the centre of the country, extending south to Kerry and Cork, and north to Antrim. It favours rich fens, and since this type of habitat is everywhere being destroyed by drainage, it is a decreasing species.

Recent work by R. H. Roberts has shown that plants from Welsh populations have characteristics which agree well with those of plants from Alpine sources as recorded by H. R. Reinhard.

Where it grows in fens, the Irish Marsh-orchid flourishes best at the edges of the main reed beds, where there is less competition, and often reappears in areas where the reeds have been cut back. The ground where it grows is always saturated, the other marsh-orchids preferring a slightly drier situation on the edges of the wetter areas. There it is accompanied by Bogbean (*Menyanthes trifoliata*), Great Sundew (*Drosera anglica*), Early Marsh-orchid, and Southern Marsh-orchid, although it is usually starting to fade before the latter two species come into flower, being at its best in mid-May.

It shows a strong inclination to hybridize, hybrids having been recorded with the following species—Common Spotted-orchid, Heath Spotted-orchid, Early Marsh-orchid, and Southern Marsh-orchid.

47a Lapland Marsh Orchid

Dactylorhiza lapponica (Laest. ex Hartman) Soó

See Postscript, pp. 141–3.

48 Man Orchid

Aceras anthropophorum (L.) Aitken. f.

The Man Orchid is well named since the flowers look like tiny human figures dangling all round the flowering spike, the flowers having a superficial resemblance to those of the Military Orchid.

There are usually two ovoid tubers: the one which supplies the flower spike is shrivelled, while the other which will nourish next year's plant is fat and white. Above these are a few short, thick roots.

The stem is stout, 15 to 40 cm high, and the membranous bracts are shorter than the ovaries.

At the base of the stem are three to four broad, blunt, bluish-green leaves, which are strongly veined. The lower leaves are crowded and their ends often appear scorched. The upper leaves are narrow and more pointed, clasping the stem and grading into the bracts of the lower flowers.

The flowering spike is long and cylindrical, bearing as many as ninety yellowish flowers, which may be tinged with reddish brown, and are arranged up the stem in an ill-defined spiral. The three sepals and the two upper petals form a short compact hood, the sepals often marked with a prominent red-brown border. The upper two petals are completely covered by the sepals.

The labellum is in the shape of a hanging human figure, and gives the species its name. The two narrow pointed side lobes form the arms, and the two pointed ventral lobes form the legs. Some flowers may show a small median tooth between the ventral lobes. The labellum is yellowish, while in some plants the arms and legs of the 'man' are suffused with a foxy-brown colour.

There is no spur, a distinguishing feature from the genus *Orchis*, but at the base of the labellum there is a shallow nectar-secreting pit. Ants and hoverflies have been seen to visit the flowers, presumably for the nectar, but have not been seen to remove the pollinia, whose viscidia lie close together and are covered by a bursicle. However, seed is set in a fairly high proportion of flowers.

The Man Orchid has a distinctly south-eastern distribution, from Lincolnshire in the north to Somerset in the west. It is nowhere a common plant, but is probably more numerous in Kent than in any other county. It is also well distributed in Surrey but it is strangely rare in Sussex, where it just survives in two downland sites. A third site in stabilized shingle near Eastbourne had some flourishing, strongly red-coloured flowers, but was unfortunately destroyed by building activity. It still grows in Warwickshire and the Isle of Wight.

The Man Orchid is a plant of old chalk pastures, where it seems to favour the steep slopes with their miniature terraces formed by the sheep tracks and weathering. V. S. Summerhayes felt that plants occurred most frequently at the foot of these slopes, but in fact they may be found all over them and on the flatter areas as well.

The Man Orchid is also to be found in old quarries on chalk and limestone, and can grow successfully under moderately dense scrub. I have even found it growing with Lady Orchids in a dense Hazel coppice, but in that case other Man Orchids flowered at the edge of a near-by field.

On the downs it is frequently accompanied by Common Milkwort (*Polygala vulgaris*) and Chalk Milkwort and by Fragrant Orchid and Pyramidal Orchid.

The flowering period is long, from early May well into July.

Abnormalities are not common, but I have found a plant whose flowers had a double hood, labellum, and column, but only one ovary.

The hybrid with the Monkey Orchid occurred in Kent in 1985.

49 Pyramidal Orchid

Anacamptis pyramidalis (L.) Richard

[*Orchis pyramidalis* L.]

The Pyramidal Orchid has the great virtues that accompany a robust constitution—it is an early and successful colonizer of waste ground on calcareous soils, and where it does become established it often flowers in profusion, the bright pink flowers making a splendid show. It is also remarkably persistent, still flowering today on chalk downland near Sevenoaks in Kent where it was already well known when recorded in 1871.

Even in recent years it has shown its ability to spread to new areas. For example, in North Wales in the late 1950s it appeared at Ynyslas on the south side of the Dovey and at Dyffryn near Harlech. In 1967 it flowered at Tywyn and by 1976 had spread to Aberdovey, while the plants at Dyffryn had multiplied and plants had flowered in Harlech. In the Hebrides, although it was known from the Outer Isles and from Coll, in 1973 I found it for the first time in Tiree.

The Pyramidal Orchid has two round tubers, crowned with a few slender fleshy roots, and a slender stem 20 to 60 cm high.

The three to four basal leaves are narrow and pointed. They develop in the autumn and over winter, often being shrivelled by the time the plant flowers. There are about six narrow sheathing stem leaves, the upper ones pointed, lanceolate, and resembling the lower bracts, which are longer than the ovaries.

The closely packed pink flower spike is shaped like a pyramid when the flowers first open, and gives the species its name, although when the upper flowers in the spike are fully open its shape may be more spherical. The colour is usually bright pink, but flowers may be found ranging in colour from pale pink to deep red, the latter colour being more often found near the sea. Albinism is not particularly common, and I have only found white flowers on four occasions in the last twenty-six years. They are strikingly beautiful when they do occur.

The two lateral sepals are oblong and spreading, parallel to the lateral lobes of the labellum, and of the same length. The upper sepal and two upper petals are small, rounded, and form a loose hood. The labellum has three well-divided equal lobes, which are rounded and spreading, the base bearing two prominent upright pink plates which are angled in to the mouth of the spur and act as a guide to visiting insects. The spur is very long, thick, and straight, projecting across to the opposite side of the flower spike. It does not contain free nectar, but there is a sweet fluid in the spur wall which attracts visiting insects, usually butterflies and moths, with the occasional *Diptera*. The flowers have a faint sweet scent.

The mechanism of pollination is highly developed and has already been mentioned in the section on fertilization (page 11). The two pollinia are borne on a single saddle-shaped viscidium which is especially well adapted to clamp on to the insect's proboscis. The pollinia then swivel downwards and outwards to contact the laterally placed stigma of the next flower visited.

Seed is set in 65 to 95 per cent of capsules, an indication of the efficiency of this method of fertilization. After germination the young plant is entirely dependent on its mycorrhizal fungus, the first leaves developing in the fifth year. Several further years elapse before the plant flowers.

The Pyramidal Orchid is widely distributed in south England, the Midlands, Wales, the Isle of Man, and the western seaboard of Scotland. It also occurs throughout Ireland. It is particularly common in dry pastures on chalk and limestone, often in the taller grasslands composed of *Bromus erectus*, *Helictotrichon* spp., and *Arrhenatherum elatius*. In western areas it is fairly frequent in calcareous sand-dunes. It is accompanied by Musk Orchid, Fragrant Orchid, Bee Orchid, Late Spider-orchid, Lizard Orchid, Monkey Orchid, and Man Orchid.

The flowering period is usually several weeks later than the Fragrant Orchid, from mid-June to mid-August.

Apart from the colour variations and albinos already noted, there is a form called *emarginatus* in which the labellum is entire and not lobed. This was reported to me by A. G. Hoare from chalk downland in Sussex in 1973. It has also been recorded from Selborne in Hampshire and in calcareous sand-dunes in Ireland.

The hybrid with the Fragrant Orchid has been recorded, but only infrequently.

Postscript to the New Edition

Since the publication of the first edition of this book in 1980, two species, Young's Helleborine (*Epipactis youngiana*) and Lapland Marsh-orchid (*Dactylorhiza lapponica*), have been found new to the British Isles. These are described in the following postscript and illustrated in the colour plates section.

9a Young's Helleborine
Epipactis youngiana A. J. Richards & A. F. Porter sp. nova

A. J. Richards, A. F. Porter, and G. A. Swan have for some years been making a detailed study of the genus *Epipactis* in Northumberland, discovering interesting populations of *E. leptochila*, *E. dunensis*, and *E. phyllanthes* on zinc- and lead-polluted soils by the River South Tyne. In the course of these studies they found, in 1976 and 1977, two populations of helleborines which could not be assigned to any known taxon. It was necessary to create a new name, which was chosen to commemorate the late D. P. Young, who had done much excellent research in the genus *Epipactis*.

The plants of the first colony grow in an oakwood on clay soil, with a dense carpet of bramble. Those of the second colony grow on a soil polluted by lead and zinc, the site being an abandoned lead mine, with planted pine trees and naturally occurring alder, sallow, and birch.

In full sun *Epipactis youngiana* is a robust plant and resembles the Broad-leaved Helleborine (*Epipactis helleborine*) in stature, but it can quickly be distinguished by its yellower foliage, and four to seven narrower, wavy-margined leaves, carried in two ranks on solitary slender stems 30 to 58 cm high. The upper part of the stem, from the highest leaf to the tip, is downy, while the lower stem is smooth.

The flower spike tends to be one-sided, bearing a dozen or more large, broadly campanulate flowers, drooping on moderately long pedicels. The bracts are long and the ovaries glabrous. The lateral sepals are long and outwardly curved. The upper petals and the epichile of the labellum are rose-flushed, a colour distinct from the reddish suffusion often seen in *E. helleborine*. The hypochile is spotted inside with purple, and the heart-shaped epichile is strongly reflexed, like that of *E. dunensis*. The two caruncles at the base of the epichile are well marked, rough, and confluent.

The viscidium (sticky cap to the rostellum) withers early, so that the pollinia crumble on to the stigma and self-pollination occurs. Although usually autogamous, I have found a few flowers where the pollinia had been removed intact. This is an important distinction from *E. helleborine*, where the viscidium persists until removed by a visiting insect.

The stigma has three sharp, forwardly directed points, the central upper point formed by the rostellum. These pronounced points are distinctive and diagnostic.

The wasp *Vespa germanica* has been observed visiting flowers, but was not seen to remove pollinia. Ants have been seen drinking the secretion from the hypochile.

Epipactis youngiana flowers during July, slightly earlier than *E. helleborine*.

Discussions at an early stage in the investigation of this helleborine led to a premature comment in the first edition of this book (p. 143) that *Epipactis confusa* had been identified. Further study showed that *E. confusa* has smaller, greener, drooping flowers which do not open so widely. Flowers of the closely related *Epipactis muelleri* lack the strongly three-pointed stigma, and the plants are less robust.

In the sites where *E. youngiana* grows it is accompanied by *E. helleborine*, *E. phyllanthes* var. *pendula*, *Dactylorhiza purpurella*, and *D. fuchsii*. The floral characters suggest that it may have originated as a hybrid between *E. helleborine* and *E. phyllanthes* var. *pendula*.

47a Lapland Marsh-orchid

Dactylorhiza lapponica (Laest. ex Hartman) Soó

The publication in 1988 of a joint paper by A. G. Kenneth, M. R. Lowe, and D. J. Tennant describing this new orchid for the British Isles followed many years of painstaking study of tetraploid marsh-orchid populations in Scotland.

They discovered a number of marsh-orchid populations in Kintyre, Ardnamurchan, Morvern, and South Harris which would not fit into any known British taxon, and which were confined to a distinctive habitat—base-rich hill flushes at low altitude.

The plants vary in height from 6 to 24 cm, rather slender, the upper stem often suffused with purple. There are two to three sheathing leaves and up to two non-sheathing leaves, pale green and more or less uniformly covered on the upper surface with dark violet brown dots, rings, or blotches. The bracts are fairly large, often purple-tinged, and always spotted on one or both surfaces.

The flowers are magenta purple or magenta red, with dark markings. The lateral sepals are erect, blunt-ended, and always marked with dark dots, spots, or rings. The upper sepal and the two upper petals form a loose helmet. The labellum is flat or with the lateral lobes slightly reflexed, having a longer broad-based central lobe.

The markings are intensely dark violet purple, often merged into a central dark patch. The spur is broad, more or less cylindrical, straight, and blunt-ended.

The flowering period commences in late May, depending on the season's weather, and lasts into July.

Although Lapland Marsh-orchid grows in base-rich flushes, it seems capable of tolerating the more acidic conditions on nearby wet heath.

The Scottish plants of *D. lapponica* agree well in all their characteristics with those recorded from northern Scandinavia and the European Alps. The species is recorded from continental sites in Switzerland, Austria, and Italy.

The following key to the tetraploid marsh-orchids of Scotland was designed by Mr Lowe, and I am most grateful for his permission to print it.

KEY TO SCOTTISH TETRAPLOID MARSH-ORCHIDS

1. Plants usually with more than 20 flowers in dense spike; more than 6 leaves. Flowering June–July 2
1. Plants with less than 20 flowers in ± lax or dense spike; usually less than 6 leaves. Flowering May–June 3
2. Leaves spotted throughout upper surface; bracts spotted ....
D. purpurella subsp. *majaliformis*
2. Leaves unspotted or few small spots towards apex; bracts unspotted
D. purpurella subsp. *purpurella*
3. Leaves crowded at base; spike dense; labellum distinctly three lobed..................................... *D. majalis* subsp. *scotica*
3. Leaves well spaced; spike ± lax; labellum three lobed or rhombic 4
4. Lower leaves oblong-obovate, usually dull green and heavily spotted; labellum usually purple with red or purple markings......
D. lapponica
4. Lower leaves linear-lanceolate, pale green, usually unmarked or lightly spotted; labellum magenta-purple with purple markings ....
D. traunsteineri

Hybrids

This section gives a comprehensive list of all the known orchid hybrids which have occurred in Great Britain, with brief descriptive notes where possible. Hybrids are identified by the scientific names of the two parent species written in alphabetical order, and the list is also arranged alphabetically to facilitate the finding of the required description in the text.

Because hybrids are by their very nature subject to variability, these notes can only serve as a guide-line, which must be supplemented by personal experience in the field.

Aceras anthropophorum × *Orchis simia*
Recorded in Kent in 1985, this being the first record for the hybrid in the British Isles although it is well known from the Continent, particularly from France. Characteristics of both parents were evident in the flowers, which had a greenish hood and small spur like *Aceras*, while the labellum had long 'arms' and a 'tail' like *O. simia*.

Anacamptis pyramidalis × *Gymnadenia conopsea*
This hybrid is possible, since pollination by moths and butterflies could occur as the flowering periods overlap. Recorded in Hampshire, Gloucestershire, and Durham. Flower spike long and scented. Labellum shows the two converging plates at the base as in *Anacamptis*.

Cephalanthera damasonium × *C. longifolia*
First recorded in Hampshire, May 1974. Leaves broad and ridged, flowers few in number borne parallel to stem on long stalks, pure white. Long perianth segments slightly spreading. Labellum with three orange ridges. Ovary intermediate in thickness, but showing a twist through 180° anticlockwise. Recorded also in France and Germany.

Coeloglossum viride × *Dactylorhiza fuchsii*
Tends to be smaller than *D. fuchsii*, the flowers pink with a suffusion of green and a curious mottled appearance. Perianth members in a loose hood, spotted labellum longer than broad, tip three lobed with small median lobe. Spur short and down curved. Most frequent on chalk downs. It was found in Co. Down (H. 38) by J. Wilde in 1986.

Coeloglossum viride × *Dactylorhiza majalis*
Reported on Harris by J. Heslop-Harrison

Coeloglossum viride × *Dactylorhiza praetermissa*
Photographs of this supposed hybrid taken by E. J. Bedford from a specimen sent from Hampshire by Col. Godfery in 1917 are at the British Museum (Natural History). They are unnumbered.

Coeloglossum viride × *Dactylorhiza purpurella*
Recorded on Iona, Eigg, and Harris, also in Co. Durham. Small plants about 11 cm high, with erect light-green spotted or unspotted leaves. Labellum intermediate with a small median lobe, purple tinged with green and marked with spots and loops.

Coeloglossum viride × *Gymnadenia conopsea*
Recorded in well-scattered localities throughout Great Britain, usually sterile. Flowers resemble *Gymnadenia* more closely, with an intermediate green-tinged labellum and a short spur.

Coeloglossum viride × *Platanthera bifolia*
Reported in 1949 in South Uist by J. Heslop-Harrison (*Vasculum* 34, 22) but not confirmed.

Dactylorhiza fuchsii × *D. incarnata*
This hybrid has been recorded in twelve vice-counties. It is taller than *D. incarnata*, leaves pale green, hooded, and sometimes spotted. The flowers vary in colour, but usually show the flesh-pink base colour of *D. incarnata*. Labellum flat and three lobed, with a pattern of broken loops and dots, spur broad and conical. Hybrids involving the subspecies *cruenta* have been recorded from the south end of Lough Carra, East Mayo, where *D. incarnata* ssp. *cruenta* grows with the *Schoenetum* in the highly calcareous peat near the lake shore, and *D. fuchsii* on the drier ground near by. The flowers are dark purple with the lateral lobes reflexed. Spur intermediate. Bracts and upper leaves heavily blotched on both surfaces.

Dactylorhiza fuchsii × *D. maculata* ssp. *ericetorum*
The hybrid has the robust growth of *D. fuchsii*, but narrower lower leaves which are not so blunt at the tip. Flowers heavily marked, labellum three lobed with broader more crenate lobes than *D. fuchsii*. Spur not so slender as in *D. maculata* ssp. *ericetorum*. The hybrid is

highly sterile. It has a scattered but widespread distribution where the acid/base-soil conditions allow the parent plants to flower in close proximity.

Dactylorhiza fuchsii × *D. majalis*

A robust hybrid with heavily spotted leaves, spots occasionally ring shaped. Flowers intermediate.

Dactylorhiza fuchsii × *D. praetermissa*

The hybrid is widely distributed and abundant where both parents occur. Leaves spotted, sometimes heavily, stem tall and hollow, flower spike dense and conical. The bracts may be longer than the flowers but are narrower than those of *D. praetermissa*. Labellum shows the loops and dots typical of *D. fuchsii*. Spur thick and broad.

Dactylorhiza fuchsii × *D. purpurella*

An article on a hybrid swarm of this parentage was published in *Watsonia*, 11 (1977), 205–10. They proved to be very variable in form, showing both spotted and unspotted leaves and intermediate flowers.

Dactylorhiza fuchsii × *D. traunsteineri*

A small plant with four to five slender, pale green, faintly spotted leaves. Labellum three lobed, lateral lobes slightly reflexed. Pale pink flowers fairly heavily marked with red loops and dots. Outer perianth segments long and pointed. Spur broad, pointed at the tip and slightly down curved. It is recorded from Anglesey and North Yorkshire, while I found two plants in a fen near Dereham in Norfolk in early June 1977. Additional records from Co. Westmeath and East Mayo.

Dactylorhiza fuchsii × *Gymnadenia conopsea*

Not an uncommon hybrid on downland especially in south England, but recorded also in the Hebrides and in Ireland. Leaves long and narrow like *Gymnadenia*, but lightly spotted. Tall compact spike. Labellum shaped like *Gymnadenia* but marked with loops and spots. Spur very long and decurved, flowers scented.

Dactylorhiza fuchsii × *Platanthera bifolia*

Photographs of this supposed hybrid taken by E. J. Bedford of a plant from an unknown site are at the British Museum (Nat. Hist.). They are numbered by E. J. B. as E851 and E852.

Dactylorhiza incarnata × *D. maculata* ssp. *ericetorum*

Stem slightly hollow, leaves yellowish green with large pale spots, upper leaves bract-like. Broad rounded labellum, lateral lobes crenate with a small median tooth. Colour ranges from the typical bright pink of *D. incarnata* to reddish purple, marked with dots in parallel loops. Spur wide throated but slender and curved. The F1 hybrid is sterile and no seed is set. This hybrid is rare, but has been recorded from thirteen vice-counties in Great Britain and two in Ireland.

Dactylorhiza incarnata × *D. majalis*

Plants are more slender than *D. incarnata*, with broad, yellow-green leaves which may be spotted. Flowers vary in colour from pale pink to red purple, and are larger and brighter than those of *D. incarnata*. Labellum intermediate, with reflexed lateral lobes. This hybrid has been recorded in Limerick and in the Hebrides.

Dactylorhiza incarnata × *D. praetermissa*

Plants resemble *D. praetermissa* more closely, with a hollow stem and erect, yellow-green, more or less hooded leaves. Bracts long. Flowers range from pale to bright lilac, depending on the colour form of *D. incarnata* involved. Labellum rather narrow, three lobed with the lateral lobes reflexed and strongly marked. Spur short and conical. The hybrid is rare but scattered in eighteen vice-counties as far north as South Lancashire and west to Pembrokeshire.

Dactylorhiza incarnata × *D. purpurella*

Plants intermediate in form, with narrow leaves which may be spotted. Flowers often dark coloured, marked with small spots and broken lines. The hybrid is rare, but has occurred in marshes and dune slacks in north England, north-west Wales, and parts of Scotland.

Dactylorhiza incarnata × *D. traunsteineri*

The hybrid is like *D. traunsteineri* in growth, with narrow, lanceolate leaves, yellow green in colour and marked with a few small spots. Upper leaves exceed the base of the flower spike, while the bracts are long and suffused with purple. Flowers are intermediate in form. It is a very rare hybrid, and has been recorded from Caernarvon, Anglesey, and Wicklow. Additional records Mid-west Yorkshire 1977 and North-east Yorkshire near Pickering 1982.

Dactylorhiza incarnata ssp. pulchella × D. maculata ssp. ericetorum
Additional record West Cornwall 1974.

Dactylorhiza incarnata × Gymnadenia conopsea
Found in West Cornwall 1984.

Dactylorhiza maculata ssp. ericetorum × D. majalis
The hybrid is taller than *D. majalis* with narrow spotted leaves. Labellum broad and flat marked with broken lines and dots, lateral lobes slightly crenated. The hybrid is very fertile. Recorded in Ireland in Kerry and West Mayo. Additional Irish record in West Clare.

Dactylorhiza maculata ssp. ericetorum × D. praetermissa
Plants resemble the hybrid *D. fuchsii* x *D. praetermissa*, but the leaves are longer, narrower, and more erect. Labellum has three shallow lobes, lateral lobes crenate, marked all over with a pattern of lines and dots. Spur long and straight, of intermediate thickness. The distribution of this hybrid is not clear.

Dactylorhiza maculata ssp. ericetorum × D. purpurella
Very variable, long narrow leaves, spotted or unspotted. Flowers vary from pale lilac to dull purple, with spreading lateral sepals. Labellum rounded, with crenate lateral lobes, marked with a pattern of purple spots and broken lines. Abundant and widespread where *D. purpurella* occurs.

Dactylorhiza maculata ssp. ericetorum × D. traunsteineri
Recorded in Caernarvon and Anglesey where *D. traunsteineri* grows in the fens and *D. maculata* ssp. *ericetorum* on the drier acid ground above. Stem slender and flexuous, bearing four to six narrow lanceolate spotted leaves. Bracts longer and broader than in *D. maculata*. Lilac-coloured flowers, large three-lobed labellum with long central lobe. Lateral lobes broad, crenated, and reflexed. Labellum marked with reddish flecks and dots. Additional records in Mid-west Yorkshire 1976, North-east Yorkshire 1975; in Ireland in Carlow, Dublin, and West Mayo.

Dactylorhiza maculata ssp. ericetorum × Gymnadenia conopsea
Not uncommon in the north and west. C. A. Stace recorded it at the Spittal of Glenshee, while I have seen it in Wester Ross and in

Merioneth. Leaves fine, narrow, and spotted. Scented flower spike resembles *Gymnadenia*, the bracts long and pointed projecting well beyond the unopened buds. Flower shape resembles *Gymnadenia*, but lateral sepals and labellum marked with fine spots. Spur fairly stout, straight, and at least twice as long as the ovary. Additional record in South-east Galway.

Dactylorhiza maculata ssp. *ericetorum* × *Platanthera bifolia*
Photographs of this supposed hybrid taken by E. J. Bedford on Ashdown Forest, Sussex, 20.6.31. Photographs at the British Museum (Nat. Hist.) are numbered by E. J. B. as E848.

Dactylorhiza maculata ssp. *ericetorum* × *Pseudorchis albida*
Recorded near Stenness, Orkney, in 1977, this being the first record for the hybrid in the British Isles. Stem and leaves more like *D. maculata*, but flowers intermediate. Floral segments spotted. Lip short, rectangular with small side lobes. Other segments converge to form a hood. Spur intermediate.

Dactylorhiza majalis × *D. purpurella*
Hybrid intermediate in form and fertile. In west Ireland and in north-west Scotland apparent hybrids occur in extensive colonies. However, in Anglesey, where both species are found, they stay distinct without intergrading.

Dactylorhiza praetermissa × *D. purpurella*
This hybrid is rare since the species scarcely overlap in distribution, but has been reported from Harlech, Merioneth, by P. M. Benoit, and is known from Cardigan. Plants at Harlech had typical broad unspotted leaves of *D. praetermissa* but the square-topped, squat inflorescence of *D. purpurella*. Flowers pale pink with a broad, scarcely lobed labellum, marked with reddish splodges especially in the central area.

Dactylorhiza praetermissa × *D. traunsteineri*
Records of this hybrid come from West Norfolk and Cambridgeshire. Hybrid intermediate in form and fertile, backcrossing resulting in a range of types. Tend to be more robust than *D. traunsteineri*, which they replace where the fens are drying out.

Dactylorhiza praetermissa × Gymnadenia conopsea

Robust plants, some with spotted leaves. Flowers resemble *D. praetermissa* but labellum markedly three lobed. Bracts large. Spur slender, shorter than *Gymnadenia*.

Dactylorhiza purpurella × Gymnadenia conopsea

Reported in Teesdale and in Scotland. Additional record in Banffshire in 1985.

Epipactis atrorubens × E. helleborine

Recorded in Denbighshire, Mid-west Yorkshire, and West Sutherland. Characteristics intermediate. The sepal lengths of pure *E. atrorubens* can vary very greatly, plants in Teesdale and North Wales having short blunt sepals, while others in Wester Ross had long pointed sepals. Similarly, flowers may be green tinged or even cream coloured, so that great care must be taken in identifying possible hybrids.

Epipactis helleborine × E. leptochila

Reported by M. J. Godfery in Surrey and Gloucestershire but not confirmed.

Epipactis helleborine × E. purpurata

Possible hybrids of very variable form recorded in south England and south Midlands.

Gymnadenia conopsea × Pseudorchis albida

Recorded in several parts of Scotland, especially in Inverness, and also in Yorkshire. V. S. Summerhayes describes the hybrid as intermediate and variable, plants tending to resemble *P. albida*, but pink in colour with a white labellum. Additional records in Perth and Easterness.

Ophrys apifera × O. fuciflora

Reported from East Kent. C. B. Tahourdin records three plants of this hybrid in flower in 1924 near Folkestone.

Ophrys apifera × O. insectifera

A striking hybrid which has been recorded in woods near Bristol since 1968. The sepals are larger than those of *O. insectifera* but held in a similar fashion. They are yellow green, sometimes with a faint pink flush. The upper petals resemble those of *O. insectifera*, the labellum broad and convex like that of *O. apifera*, the apex shallowly three-lobed, the ground colour dark brownish purple with two conjoined blue patches near the base.

Ophrys insectifera × O. sphegodes
Recorded in East Kent. I found two plants of this hybrid near Wye, Kent, in May 1960. Sepals shaped like those of *O. sphegodes*. Upper petals dark brown and slender, like the 'antennae' of *O. insectifera*. Labellum brown, very long and narrow, almost as narrow as the normal *O. insectifera*, but without lateral lobes.

Orchis mascula × O. morio
Reported from Westmorland in 1985.

Orchis militaris × O. simia
Reported in nineteenth century in the mid-Thames Valley, when both were relatively frequent. Now extremely unlikely to occur.

Platanthera bifolia × P. chlorantha
Reported in Cumberland, Kintyre, and possibly near Dolgellau in Merioneth (P. M. Benoit). It is said that abnormalities are frequent in these two species, but in my experience this is not so. The hybrid could well occur in North Wales, where both species flower in abundance in the hill hayfields.

Erratics

While the text so far has dealt with the orchid species which occur with some degree of regularity within the British Isles, there remain a small number of species whose appearance has been erratic or disputed, the plants appearing outside their normal geographical range. For this reason they have not been included as British species.

The Lax-flowered Orchid (*Orchis laxiflora*) occurs in central and south Europe and in the Channel Islands, where it grows in marshy fields. Botanically the Channel Islands may be considered as part of the Continent, so that the Lax-flowered Orchid has been excluded from this survey. However, Godfery (1933) recorded it in Durham.

The Short-spurred Fragrant Orchid (*Gymnadenia odoratissima*) has been recorded once only, many years ago, on oolitic limestone near Durham. The leaves are linear and glaucous green, the labellum three lobed like that of the ordinary Fragrant Orchid, but rather long, while the spur is broader and only as long as the ovary. The species has a mainly central European distribution, its nearest records to Durham being Scandinavia and central France.

Bertoloni's Mirror Orchid (*Ophrys bertolonii*) recorded in 1976 at Langton Matravers proved finally to be of Mediterranean origin, the soil around the plants being alien to the Purbeck area. The plants have been removed.

Also in 1976 the False Orchid (*Chamorchis alpina*) was reported from the New Forest area of Hampshire, a curious record, since in Europe this is a distinctly alpine plant, growing in damp calcareous soils between 1,000 and 2,500 metres in the Alps and in Scandinavia.

Reports of Calypso (*Calypso bulbosa*) from the north of Scotland remain unconfirmed. A visit to a number of sites in northern Sweden to see Calypso in flower showed that it grew under conditions which are unlikely to be reproduced in the area of Scotland where it was rumoured to grow. It would make an exciting addition to our flora if it were to be found truly native.

No further evidence for the existence of *Epipactis muelleri* in woodlands in the British Isles has come to light. The odd helleborines found by A. J. Richards and A. F. Porter in Northumberland in 1976 and 1977 finally proved to belong to an entirely new species, for which the name *Epipactis youngiana* was coined. *Epipactis confusa* must be removed from the ranks of plants of the British Isles.

Conclusion

There is a great temptation for all who are fascinated by wild flowers and keen that they should be conserved to wring their hands in anguish as succeeding years see further erosion of old grasslands and wetlands by agriculture, industry, and building. Certainly this is happening, and should be resisted wherever possible.

All of us who wish to see future generations enjoying the botanical delights which we possess should actively support the Nature Conservancy Council and the various Nature Conservation Trusts in the invaluable work they do in conserving and managing sites of botanical interest. Additional members of Conservation Trusts all help by their financial contribution, and for those who wish to help in a more practical way, most Trusts and many other bodies organize teams of volunteer workers for conservation jobs on nature reserves.

There are two facets of the over-all picture which give us cause for pleasure and hope for the future. In the north and west of Britain, and in the islands off the north-west coast, there has been increasing botanical activity over the last two decades, leading to the discovery of plant species new to the British Isles, and the extension of the known range of many other species. Among these are included many orchids, notably the Slender-lipped Helleborine, Pendulous-flowered Helleborine, Irish Lady's-tresses, and the Lapland Marsh-orchid.

There is always the opportunity for the amateur botanist on holiday in remote areas to discover new and interesting plants, and such information should be sent to the Botanical Recorder for the county concerned or to the BSBI c/o British Museum (Natural History), Cromwell Road, London, SW7 5BD. In this way we can build up a more comprehensive picture of the distribution of our wild flowers. Holiday-makers have a happy knack of finding unusual plants in new areas, which other botanists have passed over as unlikely to yield anything of interest.

The second cheering picture is the establishment of interesting species, including many orchids, in derelict industrial areas. In a fascinating article in *Countryman* (summer 1977), R. P. Gemmell described the orchids which have become established in areas of the

Croal–Irwell Valley north of Manchester. Owing to the deposition of alkali wastes from industrial processes a base-rich soil has gradually built up, and now supports dense and flourishing colonies of Common Twayblade, Fragrant Orchid (including impressive colonies of *G. conopsea* ssp. *densiflora*), Green-winged Orchid, Common Spotted-orchid, Early Marsh-orchid, and Northern Marsh-orchid.

Another area of waste land near Wigan was polluted by chemicals, coal refuse tips, and power-station ash deposits. Here the potentially toxic high boron levels in the waste were rapidly leached out by weathering, leaving an alkaline soil. The land was damp in places, so that Alder and Willow were able to establish, and now there are flourishing colonies of Marsh Helleborine, Early Marsh-orchid, and Southern Marsh-orchid.

Local botanists may well find such astonishingly rich orchid sites in other derelict areas, and it might prove possible to create reserves with minimal expense, allowing the orchids to flourish and possibly act as a nucleus for a new expansion.

In recent years botanical photography, along with other branches of natural history photography, has seen an almost astronomical growth rate. The standard of much of the work is high, creating valuable records and giving pleasure to many. All this activity inevitably puts an increasing burden on the environment, the traffic in human feet taking a particularly heavy toll of young orchid plants, which are not always easy to detect and avoid. In some instances it has proved necessary to prohibit access to some of our rarer orchids, in order to protect them from those of us who have the greatest interest in them. With a little goodwill it is always possible to reach a workable compromise, so that we may continue to enjoy our heritage of wild orchids without destroying it.

Rather more sinister has been the stealing of wild orchids, including some of the rarest species, from nature reserves set up to protect them. This has been done in such a way that it excludes the possibility of mistake by the uninformed. It is obviously the work of those who were well aware of what they were doing, and who had abused confidential information. Not only is such action illegal and likely to incur severe penalty, but it is very irresponsible, putting the perpetrators in the same class as those who steal the eggs of rare birds.

Significant advances have been made in the last few years in laboratory techniques used in the study of plant species and hybrids. Their application to the orchids of Great Britain may see the resolution of the confused taxonomy of *Epipactis* and *Dactylorhiza*, whose complexities seem to grow in proportion to the papers published.

Among all this new work, there is still an active part to be taken by the observant and conscientious amateur. Those who live in the south-east of England would do well to pay particular attention over the next few years to woodlands which lost trees as a result of the disastrous hurricane of October 1987. Where trees have fallen, the increased light may permit orchids to flower which have not been evident for many years.

Distribution of Species

For the purpose of recording plant species, Great Britain is divided up into vice-counties, 112 in England, Wales, and Scotland, and a further 40 in Ireland. The accompanying key gives the vice-county numbers and the names of the corresponding vice-counties. Where an orchid has been recorded in a vice-county since 1950, the vice-county number on the map is printed in bold type.

However, in the reorganization of county and regional boundaries which took place in 1974–5 many of the old names disappeared, although they are still used under the vice-county system. For this reason a second numbered key and map are provided, showing the outline of the new county and regional boundaries. It should be noted that the numbers on this second map do not correspond to the numbers on the vice-county system, and are only provided for reference.

England and Wales

1 West Cornwall (with Scilly)
2 East Cornwall
3 South Devon
4 North Devon
5 South Somerset
6 North Somerset
7 North Wiltshire
8 South Wiltshire
9 Dorset
10 Isle of Wight
11 South Hampshire
12 North Hampshire
13 West Sussex
14 East Sussex
15 East Kent
16 West Kent
17 Surrey
18 South Essex
19 North Essex
20 Hertfordshire
21 Middlesex
22 Berkshire
23 Oxfordshire
24 Buckinghamshire
25 East Suffolk
26 West Suffolk
27 East Norfolk
28 West Norfolk
29 Cambridgeshire
30 Bedfordshire
31 Huntingdonshire
32 Northamptonshire
33 East Gloucestershire
34 West Gloucestershire
35 Monmouthshire
36 Herefordshire
37 Worcestershire
38 Warwickshire
39 Staffordshire
40 Shropshire (Salop)
41 Glamorgan
42 Breconshire
43 Radnorshire
44 Carmarthenshire
45 Pembrokeshire
46 Cardiganshire
47 Montgomeryshire
48 Merionethshire
49 Caernarvonshire
50 Denbyshire
51 Flintshire
52 Anglesey
53 South Lincolnshire
54 North Lincolnshire

55 Leicestershire (with Rutland)
56 Nottinghamshire
57 Derbyshire
58 Cheshire
59 South Lancashire
60 West Lancashire
61 South-east Yorkshire
62 North-east Yorkshire
63 South-west Yorkshire
64 Mid-west Yorkshire
65 North-west Yorkshire
66 Durham
67 South Northumberland
68 North Northumberland (Cheviot)
69 Westmorland with North Lancashire
70 Cumberland
71 Isle of Man
113 Channel Isles

Scotland

72 Dumfriesshire
73 Kirkcudbrightshire
74 Wigtownshire
75 Ayrshire
76 Renfrewshire
77 Lanarkshire
78 Peebleshire
79 Selkirkshire
80 Roxburghshire
81 Berwickshire
82 East Lothian (Haddington)
83 Midlothian (Edinburgh)
84 West Lothian (Linlithgow)
85 Fifeshire (with Kinross)
86 Stirlingshire
87 West Perthshire (with Clackmannan)
88 Mid Perthshire
89 East Perthshire
90 Angus (Forfar)
91 Kincardineshire
92 South Aberdeenshire
93 North Aberdeenshire
94 Banffshire
95 Moray (Elgin)
96 East Inverness-shire (with Nairn)
97 West Inverness-shire
98 Argyll Main
99 Dunbartonshire
100 Clyde Isles
101 Kintyre

102 South Ebudes
103 Mid Ebudes
104 North Ebudes
105 West Ross
106 East Ross
107 East Sutherland
108 West Sutherland
109 Caithness
110 Outer Hebrides
111 Orkney Islands
112 Shetland Islands (Zetland)

Ireland

H. 1 South Kerry
H. 2 North Kerry
H. 3 West Cork
H. 4 Mid Cork
H. 5 East Cork
H. 6 Waterford
H. 7 South Tipperary
H. 8 Limerick
H. 9 Clare
H.10 North Tipperary
H.11 Kilkenny
H.12 Wexford
H.13 Carlow
H.14 Leix (Queen's County)
H.15 South-east Galway
H.16 West Galway
H.17 North-east Galway
H.18 Offaly (King's County)
H.19 Kildare
H.20 Wicklow
H.21 Dublin
H.22 Meath
H.23 West Meath
H.24 Longford
H.25 Roscommon
H.26 East Mayo
H.27 West Mayo
H.28 Sligo
H.29 Leitrim
H.30 Cavan
H.31 Louth
H.32 Monaghan
H.33 Fermanagh
H.34 East Donegal
H.35 West Donegal
H.36 Tyrone
H.37 Armagh
H.38 Down
H.39 Antrim
H.40 Londonderry

Vice-county boundaries

England

1 Avon
2 Bedfordshire
3 Berkshire
4 Buckinghamshire
5 Cambridgeshire
6 Cheshire
7 Cleveland
8 Cornwall
9 Cumbria
10 Derbyshire
11 Devon
12 Dorset
13 Durham
14 Essex
15 East Sussex
16 Gloucestershire
17 Greater London
18 Greater Manchester
19 Hampshire
20 Hereford and Worcester
21 Hertfordshire
22 Humberside
23 Isle of Wight
24 Kent
25 Lancashire
26 Leicestershire
27 Lincolnshire
28 Merseyside
29 Norfolk
30 North Yorkshire
31 Northamptonshire
32 Northumberland
33 Nottinghamshire
34 Oxfordshire
35 Salop
36 Somerset
37 South Yorkshire
38 Staffordshire
39 Suffolk
40 Surrey
41 Tyne and Wear
42 Warwickshire
43 West Sussex
44 West Midlands
45 West Yorkshire
46 Wiltshire

Wales

47 Clwyd
48 Dyfed
49 Gwent
50 Gwynedd

51 Mid Glamorgan
52 Powys
53 South Glamorgan
54 West Glamorgan

Scotland

55 Borders
56 Central
57 Dumfries and Galloway
58 Fife
59 Grampian
60 Highland
61 Lothian
62 Orkney
63 Shetland
64 Strathclyde
65 Tayside
66 Western Isles

160

County and regional boundaries after 1974–5

1 Lady's-slipper

Cypripedium calceolus

2 White Helleborine

Cephalanthera damasonium

3   Narrow-leaved Helleborine

Cephalanthera longifolia

4 Red Helleborine

Cephalanthera rubra

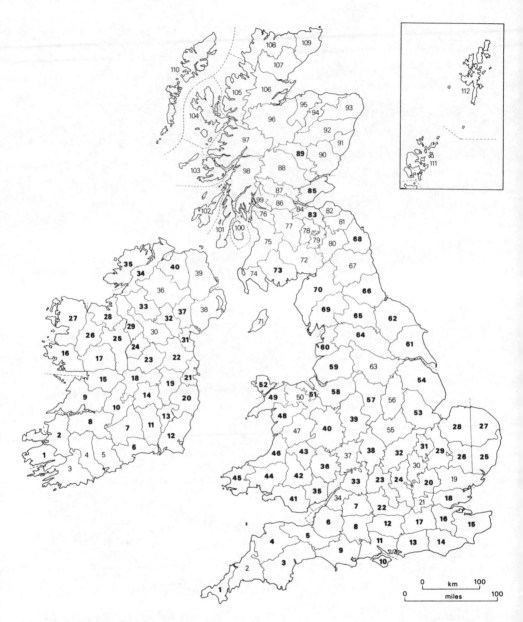

5 Marsh Helleborine

Epipactis palustris

6 Broad-leaved Helleborine

Epipactis helleborine

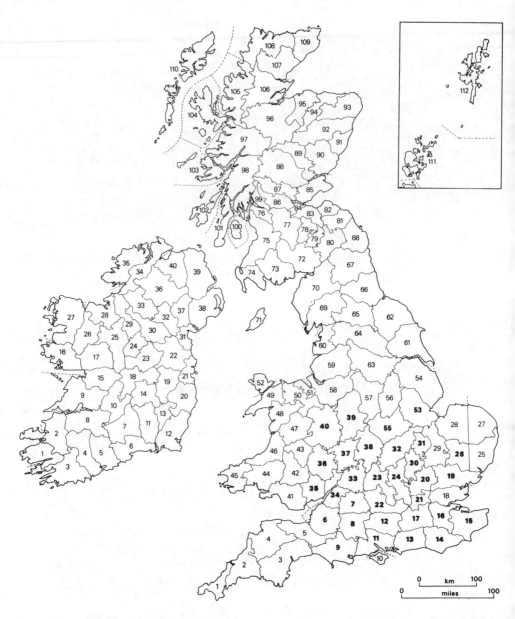

7 Violet Helleborine

Epipactis purpurata

8 Slender-lipped Helleborine

Epipactis leptochila

9 Dune Helleborine

Epipactis dunensis

9a Young's Helleborine

Epipactis youngiana

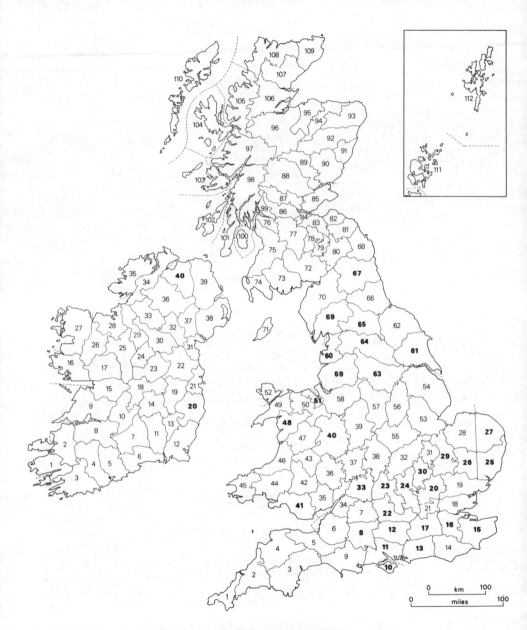

10   Pendulous-flowered Helleborine

Epipactis phyllanthes

11 Dark-red Helleborine

Epipactis atrorubens

12 Ghost Orchid

Epipogium aphyllum

13 Autumn Lady's-tresses

Spiranthes spiralis

14   Summer lady's-tresses
Spiranthes aestivalis

15 Irish Lady's-tresses

Spiranthes romanzoffiana

16 Common Twayblade

Listera ovata

17 Lesser Twayblade
Listera cordata

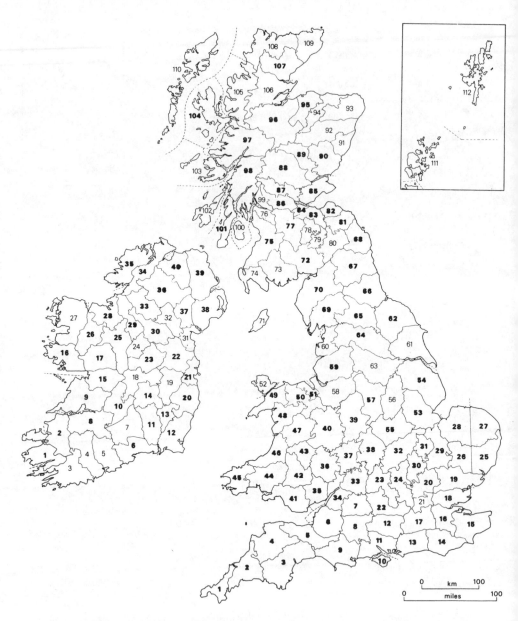

18   Bird's-nest Orchid

Neottia nidus-avis

19 Creeping Lady's-tresses

Goodyera repens

20 Bog Orchid

Hammarbya paludosa

21 Fen Orchid
Liparis loeselii

22 Coralroot Orchid
Corallorhiza trifida

23   Musk Orchid

Herminium monorchis

24 Frog Orchid

Coeloglossum viride

25 Fragrant Orchid
Gymnadenia conopsea

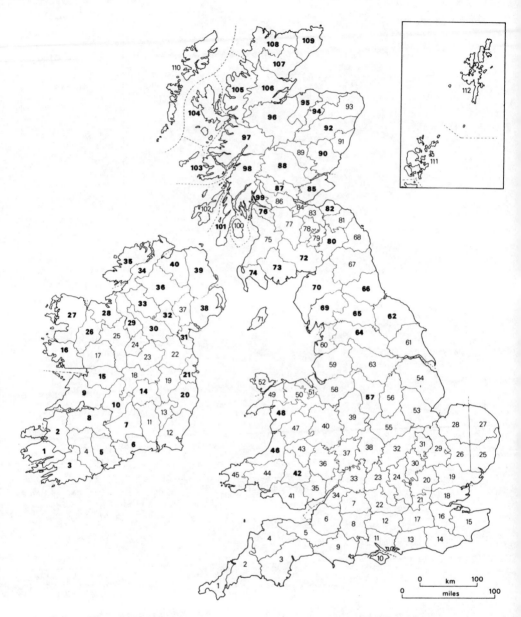

26   **Small-white Orchid**

Pseudorchis albida

27   Greater Butterfly-orchid
Platanthera chlorantha

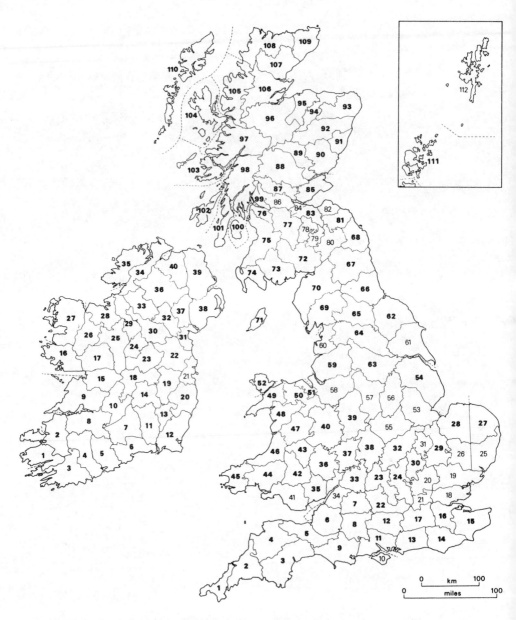

28   Lesser Butterfly-orchid

Platanthera bifolia

29   Dense-flowered Orchid

Neotinea maculata

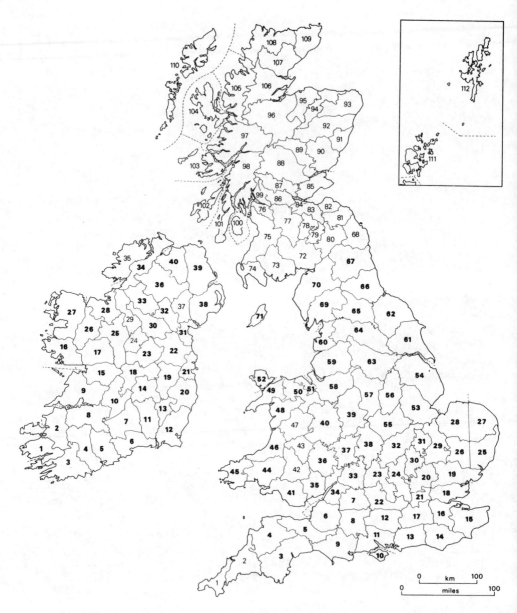

30 Bee Orchid
Ophrys apifera

31 Late Spider-orchid

Ophrys holoserica

32 Early Spider-orchid
Ophrys sphegodes

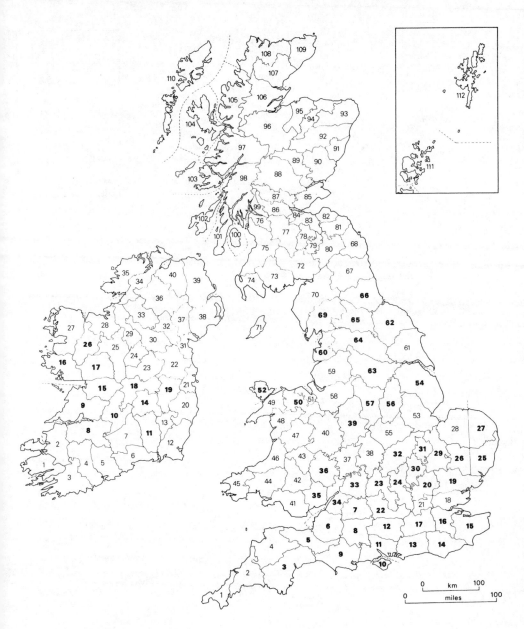

33 Fly Orchid

Ophrys insectifera

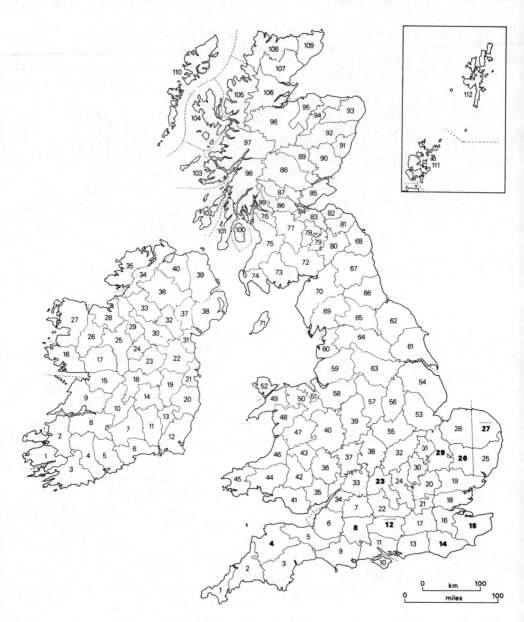

34 Lizard Orchid

Himantoglossum hircinum

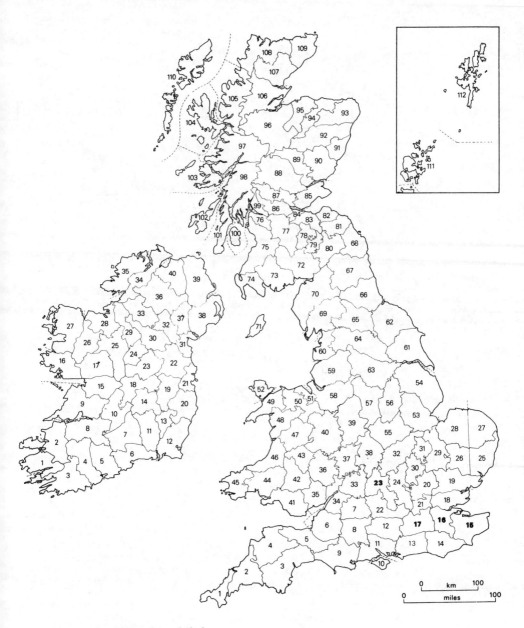

35   Lady Orchid

Orchis purpurea

36 Military Orchid

Orchis militaris

37 Monkey Orchid

Orchis simia

38 Burnt Orchid

Orchis ustulata

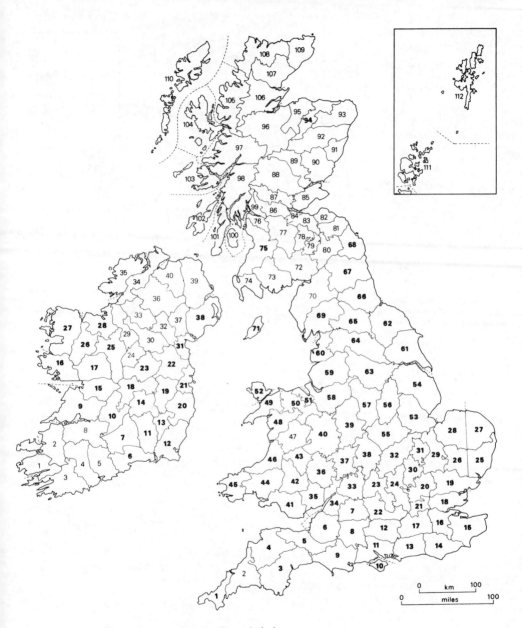

39 Green-winged Orchid
Orchis morio

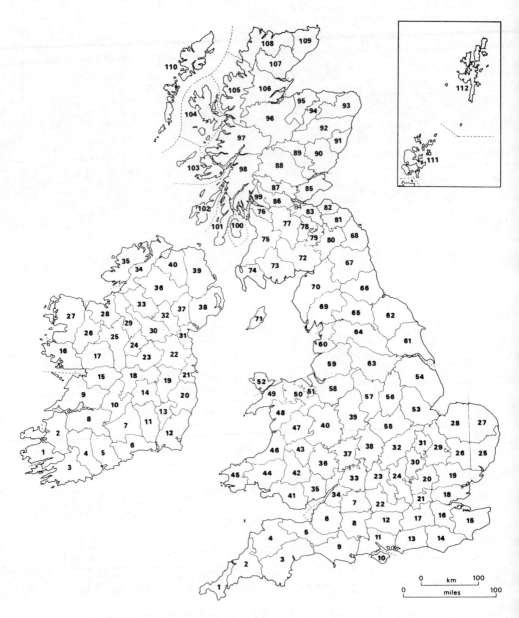

40 Early-purple Orchid
Orchis mascula

41 Common Spotted-orchid
Dactylorhiza fuchsii

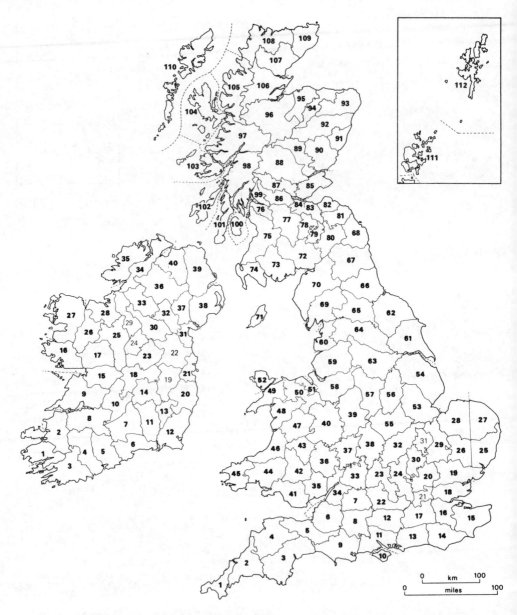

42 Heath Spotted-orchid

Dactylorhiza maculata ssp. ericetorum

43 Early Marsh-orchid

Dactylorhiza incarnata

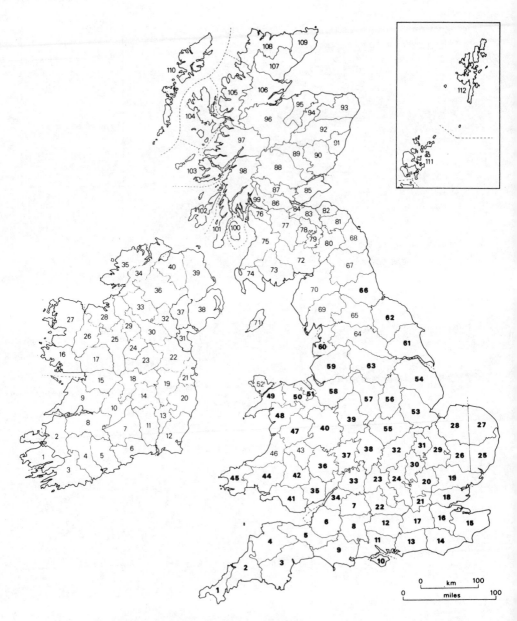

44 Southern Marsh-orchid

Dactylorhiza praetermissa

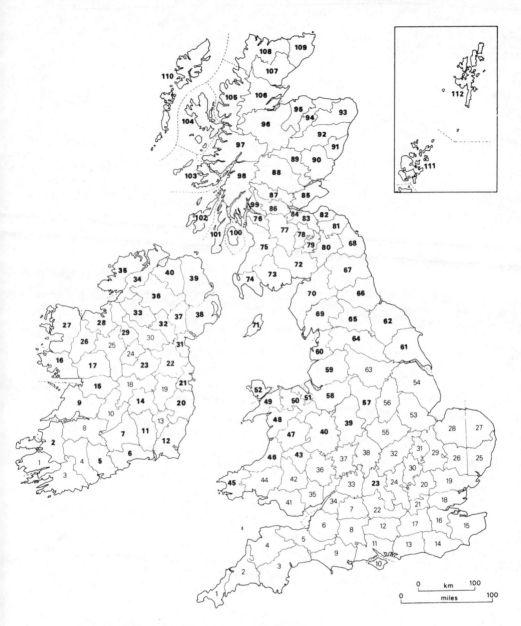

45 Northern Marsh-orchid

Dactylorhiza purpurella

46   Broad-leaved Marsh-orchid
Dactylorhiza majalis

47 Irish Marsh-orchid

Dactylorhiza traunsteineri

47a   Lapland Marsh-orchid
Dactylorhiza lapponica

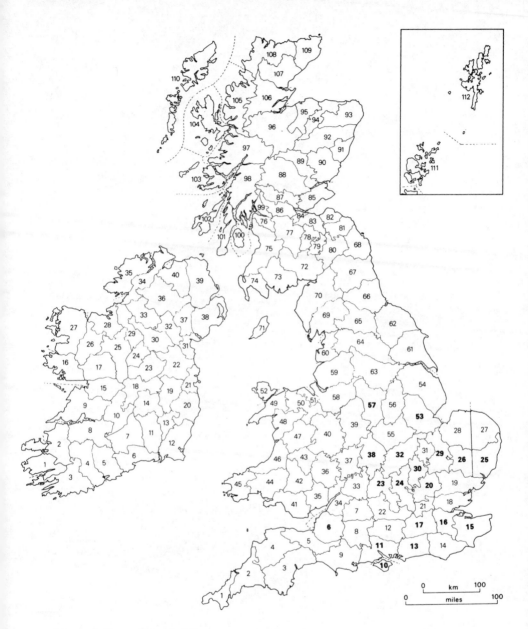

48   Man Orchid

Aceras anthropophorum

49 Pyramidal Orchid

Anacamptis pyramidalis

Bibliography

Ackerman, D. J., and Mesler, M. R. (1979) 'Pollination biology of *Listera cordata*', *Amer. Journ. Bot.* **66**: 820–4.

Ames, O., and Correll, D. S. (1943) 'Notes on American Orchids', *Bot. Mus. Harvard Univ.* **11** (1): 1–28.

Arthur, R. (1980) '*Dactylorhiza majalis* in v/c52', *Nature in Wales* **16**: 68.

Bateman, R. M. (1977) '*Epipactis helleborine*—Broad-leaved Helleborine', *J. Orchid Soc. Great Britain* **26**: 94.

(1978) '*Orchis mascula*—Early Purple Orchid', *J. Orchid Soc. Great Britain* **27**: 2.

(1978) '*Cephalanthera damasonium*—White Helleborine', *J. Orchid Soc. Great Britain* **27**: 64.

(1978) '*Epipogium aphyllum*—Ghost Orchid', *J. Orchid Soc. Great Britain* **27**: 78

(1979) '*Epipactis purpurata*—Violet Helleborine', *J. Orchid Soc. Great Britain* **28**: 2

(1979) '*Himantoglossum hircinum*—Lizard Orchid', *J. Orchid Soc. Great Britain* **28**: 60–1.

(1979) '*Neottia nidus-avis*—Bird's-nest Orchid', *J. Orchid Soc. Great Britain* **28**: 96–7.

(1980) '*Epipactis phyllanthes*—Pendulous-flowered Helleborine', *J. Orchid Soc. Great Britain* **29**: 2–3.

(1980) '*Orchis purpurea*—Lady Orchid', *J. Orchid. Soc. Great Britain* **29**: 34–5.

(1980) '*Platanthera chlorantha*—Greater Butterfly-orchid', *J. Orchid Soc. Great Britain* **29**: 67–8.

(1980) '*Cephalanthera rubra*—Red Helleborine', *J. Orchid Soc. Great Britain* **29**: 98–9.

(1981) '*Dactylorhiza incarnata*—Early Marsh-orchid', *J. Orchid Soc. Great Britain* **30**: 3–5.

(1981) '*Ophrys insectifera*—Fly Orchid', *J. Orchid Soc. Great Britain* **30**: 35–6.

(1981) '*Cypripedium calceolus*—Lady's Slipper', *J. Orchid Soc. Great Britain* **30**: 79–80.

(1981) '*Gymnadenia conopsea*—Fragrant Orchid', *J. Orchid Soc. Great Britain* **30**: 122–3.

(1982) '*Hammarbya paludosa*—Bog Orchid', *J. Orchid Soc. Great Britain* **31**: 70–1.

(1982) '*Aceras anthropophorum*—Man Orchid', *J. Orchid Soc. Great Britain* **31**: 107–8.

Bateman, R. M. (*cont.*):
> (1983) '*Orchis militaris*—Military Orchid', *J. Orchid Soc. Great Britain* **32**: 2–4.
> (1983) '*Neotinea maculata*—Dense-flowered Orchid', *J. Orchid Soc. Great Britain* **32**: 38–9.
> (1983) '*Epipactis palustris*—Marsh Helleborine', *J. Orchid Soc. Great Britain* **32**: 74–5.
> (1983) '*Dactylorhiza majalis*—Tetraploid Marsh-orchids', *J. Orchid Soc. Great Britain* **32**: 114–16.
> (1984) '*Corallorhiza trifida*—Coralroot Orchid', *J. Orchid Soc. Great Britain* **33**: 26–7.
> (1984) '*Ophrys fuciflora*—Late Spider-orchid', *J. Orchid Soc. Great Britain* **33**: 90–2.
> (1985) '*Orchis simia*—Monkey Orchid', *J. Orchid Soc. Great Britain* **34**: 88–9.
> (1985) 'Peloria and pseudopeloria in British Orchids', BSBI Meeting, *Watsonia* **15** (4): 422.
> (1986) '*Pseudorchis albida*—Small White Orchid', *J. Orchid Soc. Great Britain* **35**: 2–3.

Bateman, R. M., and Farrington, O. S. (1987) 'A morphometric study of × *Orchiaceras bergonii* (Nanteuil) Camus and its parents (*Aceras anthropophorum* (L.) Aiton f. and *Orchis simia* Lamarck) in Kent', *Watsonia* **16**: 397–407.

Baumann, H., and Künkele, S. (1982) '*Die wildwachsenden Orchideen Europas*', (Kosmos-Naturführer, Stuttgart).

Beesley, S. (1988) 'Recent Discoveries in the Irish Flora', *BSBI News* **48**: 11–12.

Benoit, P. M. (1959) '*Epipactis phyllanthes* in Merioneth', *Nature in Wales* **5**: 809–13.

Bingham, M. T. (1939) 'Orchids of Michigan', *Cranbrook Inst. Sc. Bull.*

Blackmore, S. (1985) 'Bee Orchids' (Shire Publications Ltd., Princes Risborough).

Bowen, H. J. M. (1986) 'Red Data Book plants in Berks., Bucks., and Oxon.', *Reading Nat.* **38**: 26–9.

Braithwaite, M. E. (1981) 'Report of excursion, Duns, Berwickshire', *Watsonia* **13**: 257–8.
> (1984) 'Border Pinewoods', *BSBI Scott. Newsl.* **6**: 7–9.

Breen, C. (1984) 'Report of excursion Mullingar, Co. Westmeath', *Watsonia* **15**: 175.

Brooke, J. (1950) *The Wild Orchids of Britain* (Bodley Head).

Bullard, E. R. (1985) '*Platanthera bifolia* new to Orkney', *Orkney Field Club Bull.* **2**: 9–10.

Bunce, R. G. H. (1968) 'Some New Records in Wester Ross', *Proc. Bot. Soc. Edinb.* **40**: 450–1.

Butcher, R. W. (1961) *A New Illustrated British Flora* (Leonard Hill).

Catling, P. M. (1980) '*Spiranthes romanzoffiana* or *Sp. ochroleuca?*', personal correspondence.

Clapham, A. R., Tutin, T. G., and Warburg, E. F. (1959) *Excursion Flora of the British Isles* (Cambridge University Press).

Conolly, A. (1983) 'The West Lleyn Flora', *Welsh Bulletin* **37**: 6.

Correll, D. S. (1950) *Native Orchids of North America north of Mexico* (Chronica Botanica Co., Waltham, Mass.).

Curtis, T. (1981) '*Dactylorhiza fuchsii* subsp. *hebridensis* Kincashlough Co. Donegal v/c H35', *Watsonia* **13**: 261

  (1981) '*Dactylorhiza majalis* subsp. *purpurella*. Aranmore Is., Co. Donegal v/c H35', *Watsonia* **13**: 261

Darwin, C. (1877) *The Various Contrivances by which Orchids are Fertilised by Insects* 2nd edn. (London).

Davies, P., Davies, J., and Huxley, A. (1983) *Wild Orchids of Britain and Europe* (Chatto and Windus).

Deakin, R. (1871) *The Flowering Plants of Tunbridge Wells and Neighbourhood* (Stidolph, Bellamy and Groombridge)

Dickson, J. H. (1985) '*Epipactis helleborine* in Glasgow gardens', *BSBI Scott. Newsl.* **7**: 7–8.

Donly, J. F. (1963) '*Orchids of Nova Scotia. Mill Village N.S.*' (privately printed).

Dony, J. G. (1982) 'Flowering Plants, Ferns and Fern Allies', *Bedford Nat.* **36**: 53–4.

Dony, J. G., and Dony, C. M. (1986) 'Further notes on the flora of Bedfordshire', *Watsonia* **16**: 163–72.

Doyle, G. J. (1985) 'A further occurrence of *Neotinea maculata* (Desf.) Stearn in woodland', *Irish Nat. J.* **21**: 502–3.

Duperrex, A. (1961) *Orchids of Europe* (Blandford).

Edmondson, T. (1979) '*Ophrys apifera* Huds. in artificial habitats', *Watsonia* **12**: 337–8.

Ellis, R. G. (1983) *Flowering Plants of Wales* (National Museum of Wales).

Ettlinger, D. M. T. (1979) '× *Pseudorhiza bruniana* (Brügger) P. F. Hunt in Orkney', *Watsonia* **12**: 259.

  (1987) 'Peloric and duplex examples of *Orchis purpurea* Hudson in Kent', *Watsonia* **16**: 432.

Farrell, L. (1980) 'Orchid problems', *BSBI News* **24**: 9.

  (1985) '*Orchis militaris* L.—Biological Flora of the British Isles', *J. Ecol.* **73**: 1041–53.

Foley, M. J. Y. (1986) '*Dactylorhiza maculata* (L.) Soó × *D. traunsteineri* (Sauter) Soó in N.E. Yorks.', *Watsonia* **16**: 175–6.

  (1987) 'The current distribution and abundance of *Orchis ustulata* L. in northern England', *Watsonia* **16**: 409–15.

Fowler, B. R. (1980) 'A transitional abnormality in *Platanthera chlorantha* Cust. ex Rchb.', *BSBI News* **26**: 23.

Bibliography

Gemmell, R. P. (1977) 'Wild Orchids in the Wastelands', *Countryman*, summer, 74–9.

Godfery, M. J. (1933) *Monograph and Iconograph of Native British Orchidaceae* (Cambridge University Press).

Grenfell, A. L. (1981) 'Report of visit to downland site for *Epipactis palustris* in Wiltshire', *Watsonia* 13: 253–4.

Hackney, P. (1986) '*Orchis morio* in Co. Down v/c H38. Excursion report', *BSBI News* 44: 28.

Heslop-Harrison, J. (1948) 'Field Studies in Orchis: 1. The Structure of Dactylorchid Populations in Certain Islands in the Inner and Outer Hebrides', *Trans. Bot. Soc. Edinb.* 35: 22–66.

(1949) 'Notes on the Dactylorchids of North Western Donegal', *Irish Nat. J.* 9: 291–8.

(1956) 'Some observations on *Dactylorchis incarnata* (L.) Vermln. in the British Isles', *Proc. Linn. Soc. London* 166: 51–82.

(1973) 'Flora of Coll and Tiree' (personal communication).

Hill, D. A. (1978) 'A seven year study of a colony of bee orchids (*Ophrys apifera* Hudson)', *Watsonia* 12: 162–3.

Hoare, A. G. (1980) '*Anacamptis pyramidalis* (L.) Rich. in Sussex', *BSBI News* 24: 27.

Horsman, F. (1986) '*Hammarbya paludosa* refound in Upper Teesdale' (personal communication).

Hutchings, M. J. (1987) 'The population biology of the Early Spider Orchid *Ophrys sphegodes* Mill. 1. A demographic study from 1975–1984', *J. Ecol.* 75: 711–27.

(1987) 'The population biology of the Early Spider Orchid *Ophrys sphegodes* Mill. 2. Temporal patterns in behaviour', *J. Ecol.* 75: 729–42.

Keble Martin, W. (1965) *The Concise British Flora in Colour* (Ebury Press).

Kemp, R. J. (1987) 'Reappearance of *Orchis purpurea* Hudson in Oxfordshire', *Watsonia* 16: 435–6.

Kendrick, F. M. (1983) 'Botany 1982. *Epipogium aphyllum* in Herefordshire', *Trans. Woolhope Nat. Field Club* 44: 124–5.

Kenneth, A. G., and Tennant, D. J. (1984) '*Dactylorhiza incarnata* (L.) Soó subsp. *cruenta* (O. F. Mueller) P. D. Sell in Scotland', *Watsonia* 15: 11–14.

(1987) 'Further notes on *Dactylorhiza incarnata* subsp. *cruenta* in Scotland', *Watsonia* 16: 332–4.

Kenneth, A. G., Lowe, M. R., and Tennant, D. J. (1988) '*Dactylorhiza lapponica* (Laest. ex Hartman) Soó in Scotland', *Watsonia* 17: 37–41.

Knight, J. T. H. (1977) Letter in *BSBI News* 15: 24.

Lamb, K. (1982) '*Cephalanthera longifolia* in South Tipperary (H7)', *Irish Nat. J.* 20: 454.

Lang, D. C., and Lansley, J. L. S. (1978) '*Cephalanthera damasonium* (Mill.) Druce × *C. longifolia* (L.) Fritsch.', *Watsonia* 12: 49–50.

Laurence, R. J. (1986) '*Ophrys apifera* Hudson subsp. *jurana* Ruppert found in Britain', *Watsonia* 16: 177–8.

Lord, R. M., and Richards, A. J. (1977) 'A Hybrid Swarm between the Diploid *Dactylorhiza fuchsii* and the Tetraploid *D. purpurella* in Durham', *Watsonia* 11: 205–10.

Lowe, M. R., Tennant, D. J., and Kenneth, A. G. (1986) 'The status of *Orchis francis-drucei* Wilmott', *Watsonia* 16: 178–80.

Luer, C. A. (1975) *The Native Orchids of the U.S. and Canada excluding Florida* (N.Y. Bot. Gdn.).

McClintock, D., and Fitter, R. S. R. (1956) *The Pocket Guide to Wild Flowers* (Collins).

McKean, D. R. (1982) '× *Pseudanthera breadalbanensis* McKean: A new intergeneric hybrid from Scotland', *Watsonia* 14: 129–31.

Mayr, E. (1942) *Systematics and the Origin of Species* (Columbia University Press).

Medd, T. F. (1984) '*Epipactis phyllanthes* is new to v/c 65)', *Naturalist (Yorkshire)* 1984: 149.

Morton, J. K. (1959) 'The Flora of Islay and Jura', *Proceedings of the Botanical Society of the British Isles* 3: 3.

Nature Conservancy (1968) 'Tiree and Coll Dunes and Machair', *Site Report* 37.

Nelson, E. (1979) 'Monographie und Ikonographie der Orchidaceen Gattung *Dactylorhiza*', *Taxon* 28: 592–953.

Nilsson, L. A. (1978) 'Pollination ecology of *Epipactis palustris* (Orchidaceae)', *Bot. Not. (Lund)* 131: 355–68.

—— (1979) 'Pollination ecology of *Herminium monorchis*', *Bot. Not. (Lund)* 132: 537–49.

Nilsson, S. (1977) *Nordens Orkidéer* (Wahlström and Widstrand; English edn. Penguin Books Ltd., 1979).

Nylén, B. (1984) *Orkidéer i Norden* (Natur och Kultur, Kristianstad, Sweden).

Paul, V. N. (1965) *Survey No. 1: Orchids of the Chilterns* (Chiltern Research Committee).

Peitz, E. (1970) '*Aceras-orchis* Bastarde', *Orchidée* 21: 249–55.

Perring, F. H., and Walters, S. M. (1962) *Atlas of the British Flora* (BSBI/Nelson; 2nd edn. 1976).

Pigott, C. D. (1956) 'The Vegetation of Upper Teesdale in the North Pennines', *J. Ecol.* 44: 545–86.

Povey, S. (1980) 'Abnormal *Anacamptis pyramidalis*', *BSBI News* 26: 25.

Proctor, M. C. F., and Ivimey-Cook, R. B. (1966) 'The Vegetation of the Burren', *Proc. Roy. Irish Academy* 64 b15:211–301.

Bibliography

Pryce, R. D. (1983) 'The Flora of Carmarthenshire—some aspects of the first year's recording', *Welsh Bulletin* **37**: 7–9.

Ranwell, D. S. (ed.) (1974) 'Sand Dune Machair' (National Environmental Research Council Report).

Richards, A. J. (1986) 'Cross-pollination by wasps in *Epipactis leptochila* (Godf.) Godf. s.l.,' *Watsonia* **16**: 180–2.

(1986) 'The status of *Epipactis leptochila* and *E. dunensis*', *Watsonia* **16**: 221.

Richards, A. J., and Porter, A. F. (1982) 'On the identity of a Northumberland *Epipactis*', *Watsonia* **14**: 121–8.

Richards, A. J., and Swan, G. A. (1976) '*Epipactis leptochila* and *Epipactis phyllanthes* occurring in South Northumberland on Lead and Zinc Soils', *Watsonia* **11**: 1–5.

Roberts, R. H. (1961) 'Studies on Welsh Orchids 1. The Variation of *Dactylorchis purpurella* (T. and T. A. Steph.) Vermeul. in North Wales', *Watsonia* **5**: 23–35.

(1988) 'The occurrence of *Dactylorhiza traunsteineri* (Sauter) Soó in Britain and Ireland', *Watsonia* **17**: 43–7.

Roden, C. (1979) 'A note on the flora of the Lough Corrib Region including some new records for West Galway (H16), North-east Galway (H17) and South Mayo (H26)', *Bull. Ir. Biogeog. Soc.* **3**: 34–6.

Roland, A. E., and Smith, E. C. (1969) *Flora of Nova Scotia* (N.S. Museum, Halifax, Nova Scotia).

Rose, F. (1949) '*Orchis purpurea*', *J. Ecol.* **36**: 366–77.

(1976) 'Three forms of *Gymnadenia conopsea*' (personal communication).

Rutterford, M. G. (1985) 'The lizard orchid (*Himantoglossum hircinum*) at Lakenheath—a history of happenings since 1974', *Trans. Suffolk Nat. Soc.* **21**: 50–1.

St. Christopher's School, Burnham-on-Sea (1978). 'Operation Orchid', *Watsonia* **12**: 197.

(1980) 'Operation Orchid—Disaster July 1979', *Watsonia* **13**: 168–9.

Scannell, M. J. P. (1973) '*Dactylorhiza traunsteineri* (Sauter) Soó in East Cork, Mid Cork, Offaly, Meath, Leitrim and East Mayo', *Irish Nat. J.* (**17**): 426.

(1973) *Dactylorhiza* × *Kellerana* P. F. Hunt in West Meath and East Mayo', *Irish Nat. J.* (**17**): 426.

Scannell, M. J. P., and Synnott, D. M. (1972) *Census Catalogue of the Flora of Ireland*, 107–10.

Sheppard, R., and Sheppard, E. (1985) '*Neotinea maculata* (Desf.) Stearn in County Donegal', *Irish Nat. J.* **21**: 534–5.

Sheviak, C. J. (1973) 'A new *Spiranthes* from the grasslands of Central North America', *Bot. Mus. Harvard Univ.* **23**: 285–97.

Sheviak, C. J., and Catling, P. M. (1980) 'The Identity and Status of *Spiranthes ochroleuca*', *Rhodora* **82**: No. 832.

Slack, A. A., and Stirling, A. McG. (1963) 'The Cambrian Limestone Flora of Kishorn, West Ross', *Proc. Bot. Soc. Brit. Is.* **5**: 1–12.

Smith, A. J. (1975) 'Coralroot in Selkirkshire', *Hist. Berwick. Nat. Club* **40**: 101.

Smith, G. E. (ed.) (1852) *'Epipactis phyllanthes'*, *Gardeners' Chronicle* 660.

Stace, C. A. (ed.) (1975) *Hybridization and the Flora of the British Isles* (Academic Press).

Stanley, P. D. (1984) 'Frog Orchids in Kirkcudbrightshire', *Watsonia* **15**: 171.

Steel, D., and Creed, P. (1982) *Wild Orchids of Berkshire, Buckinghamshire and Oxfordshire* (Pisces Publications, Oxford).

Stewart, O. M. (1984) 'A Scottish Miscellany', *Watsonia* **15**: 171–2.

Summerhayes, V. S. (1951) *Wild Orchids of Britain* (Collins; new edn. 1968).

Tahourdin, C. B. (1924) *Some Notes as to British Orchids* (Grubb, Croydon).

Thomas, C. (1950) 'The Kenfig *Epipactis*', *Watsonia* **1**: 283–8.

Vose, P. B., Grace Powell, H., and Spence, J. B. (1957) 'The Machair Grazings of Tiree, Inner Hebrides', *Trans. Bot. Soc. Edinb.* **37**: 89–110.

Wallis, R. C. (1983) 'Green-winged Orchids in Ayrshire', *Country-side*, n.s. **25**: 212–14.

Webb, N. P. (1979) 'Marsh orchids on an unusual site in W. Lancs.', *BSBI News* **23**: 26.

Wells, T. C. E. (1967) 'Changes in a Population of *Spiranthes spiralis* (L.) Chevall at Knocking Hoe National Nature Reserve, Bedfordshire, 1962–65', *J. Ecol.* **55**: 83–99.

(1974) 'The Flowering of some Bedfordshire Orchids', *Bedfordshire Magazine* **14**: 231–5.

(1975) 'The Floristic Composition of Chalk Grassland in Wiltshire' in *Supplement to the Flora of Wiltshire*, ed. L. F. Stearn (Wilts. Arch. N.H.S., Devizes).

Wells, T. C. E., and Farrell, L. (1984) 'Bee Orchid Survey', *Watsonia* **15**: 172–3.

Wells, T. C. E., and Kretz, R. (1986) *'Spiranthes spiralis* (L.) Chevall—from seed to flowering plant in 5 years', *Watsonia* **16**: 235.

Weston, I. (1979) 'A variant of *Dactylorhiza fuchsii* (Druce) Soó in N. Lincs.', *Watsonia* **12**: 399–400.

(1982) 'A Lincolnshire *Epipactis*—possible *E. leptochila* discovered in N. Lincolnshire v/c 54 in 1978' (personal communication).

White, J., and Doyle, G. (1978) *'Neotinea maculata (N. intacta)* in woodland', *Irish Nat. J.* **19**: 187.

Wilks, H. M. (1985) 'A new record. Hybrid *Aceras anthropophorum* × *Orchis simia'*, *Kent Trust Bull.* 1985 (3): 3.

Willems, J. H. (1982) 'Establishment and development of a population of *Orchis simia* in the Netherlands 1972–1981', *New Phytologist* **91**: 757–65.

Bibliography

Williams, J. G., Williams, A. E., and Arlott, N. (1978) *A Field Guide to the Orchids of Britain and Europe* (Collins).

Willis, A. J. (1980) '*Ophrys apifera* Huds. × *O. insectifera* L. a natural hybrid in Britain', *Watsonia* 13: 97–102.

Wilson, M. (1980) 'The flowering habits of *Ophrys apifera*', *Orchid Rev.* 88: 94–6.

Winham, J. (1981) 'Report on excursion to Cam Chreag, Mid Perthsire', *Watsonia* 13: 259.

Wolfe, E. H. (1979) '*Cephalanthera longifolia* in v/c 47', *Nature in Wales* 16: 217.

Wolley-Dod, A. H. (1937) *Flora of Sussex* (K. Saville, Hastings).

Yeates, C. S. V. (1981) 'Early flowering of *Coeloglossum viride* on magnesian limestone in N. Yorkshire' (personal communication).

Young, D. P. (1949) 'Studies in the British *Epipactis* I and II', *Watsonia* 1: 102–13.

(1952) 'Studies in the British *Epipactis* III and IV', *Watsonia* 1: 253–76.

(1962) 'Studies in the British *Epipactis* V. *Epipactis leptochila* with some notes on *E. dunensis* and *E. muelleri*', *Watsonia* 5: 127–35.

(1962) 'Studies in the British *Epipactis* VI. Some further Notes on *E. phyllanthes*', *Watsonia* 5: 136–9.

Glossary

Aberrant: a form which deviates from the normal type.

Acid: water or soils containing free acids with a pH value less than 7. Such soils lack chalk or lime.

Albinism: characteristic shown by plants which are albino.

Albino: a plant which shows a congenital lack of the usual colour pigments, normally the reds and blues. Green pigments are still present.

Alder carr: wet, fen-like areas with a strong growth of Alder or willow trees.

Alkaline: pertaining to water or soils containing lime, potash, etc., with a pH value greater than 7.

Alpine: a plant which is native to mountain regions.

Autogamous: applied to a flower which is self-pollinating.

Base rich: pertaining to a soil containing large amounts of basic substances such as compounds of calcium, potassium, or magnesium.

Bog: a wet marshy area overlying acid peat.

Bract: a small leaf-like structure at the base of the flower stalk.

Bulbil: a small bulb arising on the leaf edge, or between the leaf and stem of a plant.

Bursicle: a small flap or pouch which covers the viscidia and prevents them from drying out.

Calcareous: pertaining to water or soils containing chalk or lime.

Capsule: the seed-containing structure, composed of a number of carpels joined together, at the base of the flower.

Carboniferous: pertaining to a system of rocks formed between the Devonian and Permian periods, sometimes containing coal.

Carpel: one of the divisions of the capsule.

Carr: *see* Alder carr.

Caudicle: the stalk by which the pollinium is attached to the viscidium at its base.

Chlorophyll: the green pigment in most plant cells, which takes part in photosynthesis, the process of converting carbon dioxide and water into carbohydrates with energy from sunlight.

Chromosome: one of the basic components of the cell nucleus, which carry the inherited characters of the organism.

Cleistogamous: self-pollinating within a flower which does not fully open.

Colony: a group of plants of the same type growing in a close and well-defined area.

Column: a specialized structure in the centre of the orchid flower, formed by the flower stalk, the upper part of the female reproductive organ (stigma), and the lower part of the male reproductive organ (stamen).

Conjoined: joined together.

Connivent: pertaining to structures which are separated at their bases, but touching at their apices.

Conspecific: belonging to the same species.

Crenate: with a scalloped edge.

Dactylorchids: orchids belonging to the genus *Dactylorhiza*, having palmately divided tubers.

Dominant: pertaining to the most abundant plant form in a community.

Dune slacks: damp areas lying between the hills of the dunes.

Ectotrophic: pertaining to mycorrhizal fungi growing on the surface of the orchid roots or tubers.

Endosperm: the tissue surrounding the germinative centre of a seed.

Endotrophic: pertaining to mycorrhizal fungi growing deep in the tissues of the orchid roots or tubers.

Epichile: the outer part of the labellum of orchids of the genus *Epipactis*.

Erratic: a plant occurring in a geographical area outside its normal range.

Etiolated: applied to a plant made pale and spindly by growing without adequate sunlight.

F_1: the first-generation offspring of a hybrid.

F_2: the second-generation resulting from crossing two F_1 individuals.

Family: a group of related genera.

Fasciated: compressed together in a bundle.

Fasciculated: growing in a bunch.

Fen: a wet area where peat is overlaid by alkaline water.

Fertilization: the process of uniting male and female reproductive cells.

Flora: the plants of a particular region or environment.

Floral segment: a division of the structure of a flower, such as a sepal or petal.

Flush: a wet area on a slope, formed where water flows out from a spring.

Friable: crumbly and easily broken into pieces.

Gamete: male or female reproductive cell.

Genus: a group of related species.

Germination: the sprouting of root and shoot from a seed.

Glaucous: bluish and usually smooth.

Hood: the helmet shape formed by the connivent upper petals and sepals in certain orchid flowers.

Hooded: usually applied to the leaf tip of certain species, where the tip and edges are slightly inrolled.

Hybrid: a plant originating from the fertilization of one species by another.

Hybridization: the process by which a hybrid is formed.

Hybrid swarm: a group of hybrids which show a range of characteristics between those of the two parent plants.

Hypochile: the basal part of the labellum of orchids of the genus *Epipactis*.

Indigenous: belonging naturally to a particular region.

Inflorescence: the flowering part of the plant above the highest stem leaves.

Intergeneric: applied to hybrids formed between species of two different genera.

Intergradation: the process by which lack of distinctiveness between species permits hybridization.

Interspecific: applied to hybrids formed between plants of two different species.

Introgression: the process of repeated backcrossing of an F_1 hybrid with one of its parents.

Keeled: describing leaves which are folded along a marked midrib, producing a shape like the keel of a boat.

Labellum: the lip of the flower, in orchids the lower of the three petals, often large and complex in structure.

Lanceolate: shaped like a lance head, tapering and pointed.

Lax: loose, not closely packed.

Leached: pertaining to soil which has had the soluble elements removed by the action of water.

Ley: a temporary grassland.

Liassic: pertaining to a bluish, fossil-bearing, limestone rock of the early Jurassic period.

Limestone pavement: flat areas of limestone divided into slabs and blocks by the action of water.

Lobe: a division of a leaf, petal, or sepal, larger than a tooth, but lacking a separate stalk.

Luminescence: the property of emitting light without heat as the result of a chemical reaction.

Machair: pasture formed on stabilized sand-dunes, mainly in the Hebrides.

Marsh: a wet area overlying a soil not composed of peat.

Median: situated centrally or lying along the central axis of a structure.

Membranous: thin and papery.

Microclimate: the physical conditions present immediately around a plant.

Monocarpic: pertaining to plants which flower once only and then die.

Monocotyledons: plants which have a single cotyledon or first leaf after germination.

Monogynous: pertaining to flowers having only one female organ.

Morphology: the study of the shape and form of a plant, or that form.

Mycorrhiza: the fungus which invades the underground parts of many orchid species.

Neutral: neither acid nor alkaline, having a pH value of 7.

Oceanic: describing climate governed by the close proximity of the ocean.

Oolitic: pertaining to limestone composed of minute rounded concretions resembling fish roe, in some places altered to ironstone by replacement with iron oxide.

Ovary: the lower part of the female reproductive organ which contains the seeds.

Ovoid: egg shaped.

Palmate: divided like the fingers of a hand.

Papilla: small nipple-like projection.

Pedicel: the stalk of a single flower.

Peloric: describing a flower with regular members, while the species normally has irregular perianth members.

Perennial: a plant living for more than two years.

Perianth: the outer non-reproductive parts of the flower, divided into an outer series (sepals) and an inner series (petals).

Petal: one of the segments of the inner whorl of the perianth.

Petalloid: having a structure like a petal.

pH: logarithm of hydrogen-ion concentration in moles per litre, giving a measure of the alkalinity or acidity of a solution or soil, pH 7 being neutral, values greater than 7 being alkaline and less than 7 being acid.

Photosynthesis: the process of converting carbon dioxide and water into carbohydrates using energy from sunlight.

Pollination: transference of pollen from the male reproductive organ (stamen) to the female reproductive organ (stigma).

Pollinia: structures formed by the coherence of pollen grains into a mass.

Protocorm: the underground structure first formed when an orchid seed germinates, usually infected with mycorrhizal fungus.

Pseudobulb: a bulb-like swelling of the aerial stem, not a true bulb.

Pseudocopulation: the attempt by a male insect to mate with a flower to which it has been attracted, a process by which the pollinia are removed and transferred to another flower.

Raceme: an unbranched flower spike where the flowers are borne on pedicels.

Reflexed: folded back.

Rhizome: an underground stem, usually growing horizontally, and lasting for more than one season.

Rhizomatous: like a rhizome.

Root-stock: a short underground stem, usually growing vertically.

Rosette: a group of leaves arranged around the base of a stem like the petals of a rose, often flat on the ground.

Rostellum: the sterile third stigma of an orchid flower, situated between the stamens and the two functional stigmas. Often long and beak-shaped, bearing the viscidia of the pollinia.

Runner: a slender creeping stem at ground level or just below it.

Saccate: pouched.

Saprophytic: a plant which obtains its nutrition from the breakdown of dead plant or animal material.

Semi-peloric: describing a flower in which the abnormal perianth members give a misleading appearance of being peloric. In the semi-peloric Bee Orchid the labellum resembles the sepals. If it were truly peloric it would resemble the antenna-like upper petals.

Sepal: one of the segments of the outer whorl of the perianth.

Sessile: without a stalk.

Slacks: wet hollows in sand-dunes.

Spathe: a large bract or pair of bracts enveloping a developing flower spike.

Spathulate: paddle shaped.

Species: a group of individuals having common characteristics, a division of a genus.

Spike: an elongated unbranched flower head.

Sport: a plant deviating suddenly from the normal type.

Spur: an elongated pouch formed at the base of the labellum.

Stamen: one of the male reproductive organs.

Staminode: a sterile stamen, much altered in shape and size. Present in Lady's Slipper.

Stigma: the receptive upper part of the female reproductive organ.

Stolon: an above-ground creeping stem.

Stoloniferous: bearing stolons.

Sub-species: a division of a species, distinguished only by very slight variation, insufficient to accord it the rank of a separate species.

Swarm: a group of hybrids showing a range of characteristics between those of the parent plants.

Taxa: a division of plant classification such as a genus or a species.

Tribe: a group of plants of a rank between a genus and a sub-family.

Tuber: a swollen part of a stem or root not persisting for more than a year, tubers of successive years not arising from one another. Food stores.

Turlough: a pond or small lake which dries up in the summer (Irish).

Valve: one of the segments of a capsule which splits to allow seeds to disperse.

Variation: difference in characteristics within a species.

Variety: a division of a species showing some minor differences in character.

Vector: an agent, insect or otherwise, involved in carrying pollen from one flower to another.

Vegetative: concerned with growth and development.

Vegetative multiplication: the increase in the number of individuals by an asexual process.

Vice-county: the subdivision of a county into a smaller unit more suitable for recording species distribution.

Viscidium: the sticky disc at the base of the pollinium which glues the pollen mass on to a visiting insect.

Whorl: more than two organs of the same kind arising at the same level.

Index

A number in **bold** after the English or scientific name of an orchid species refers to the species number, colour plate number (after p. 118, and distribution map (after p. 157).

225

Index

Index

Index

Index

NOTES

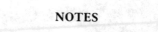

NOTES

NOTES

NOTES

NOTES

NOTES

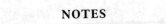

NOTES

NOTES

NOTES